THE ILLUSTRATED DICTIONARY OF

OCEANOGRAPHY

Copyright © 1993 Merlion Publishing Ltd
First published 1993 by
Merlion Publishing Ltd
2 Bellinger Close
Greenways Business Park
Chippenham
Wiltshire SN15 1BN
UK

Series editor: Merilyn Holme
Editor: Maureen Bailey

Design: Jane Brett, Steven Hulbert
Illustrations: Brian McIntyre (Ian Fleming & Associates); Jeremy
Gower (B.L. Kearley); Stephen Lings (Linden Artists); Oxford
Illustrators; Jamie Medlin (Plum Illustration); Colin Woolf (Linda
Rogers Associates); Peter Geissler (Specs Art Agency); Simon
Tegg; Fred Anderson (Bernard Thornton).
Cover illustration: Maltings Partnership

Consultant: Dr Alison Weeks, Oceanographer, working in research
at Southampton University.

Printed in Great Britain by BPCC Paulton Books Ltd

ISBN 1 85737 008 2

THE ILLUSTRATED DICTIONARY OF
OCEANOGRAPHY

Contributors
Jill Bailey
Maureen Bailey
Malcolm Tucker

Merlion Publishing

Reader's notes

The entries in this dictionary have several features to help you understand more about the word you are looking up.

- Each entry is introduced by its headword. All the headwords in the dictionary are arranged in alphabetical order.

- Each headword is followed by a part of speech to show whether the word is used as a noun, adjective, verb or prefix.

- Each entry begins with a sentence that uses the headword as its subject.

- Words that are bold in an entry are cross references. You can look them up in this dictionary to find out more information about the topic.

- The sentence in italics at the end of an entry helps you to see how the headword can be used.

- Many of the entries are illustrated. The labels on the illustrations highlight all the key points of information.

- Many of the labels on the illustrations have their own entries in the dictionary and can therefore be used as cross references.

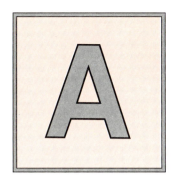

abalone　*noun*
An abalone is a large **bivalve**, which can be up to 30 centimetres long. Abalone shells have a row of holes along the outer edge. Inside the shell is a lining of **mother-of-pearl**, which is used to make jewellery. Abalone flesh is very tasty.
They went to the market to buy some abalone shells.

abyssal plain　*noun*
Abyssal plain is the name given to the large, flat area that forms much of the **ocean floor**.
The abyssal plain lies at a depth below 3,500 metres.

Adriatic Sea　*noun*
The Adriatic Sea is a **gulf** of the **Mediterranean Sea**. It is bordered by the **coasts** of Italy and Albania, Montenegro and Croatia.
The Italian coast of the Adriatic Sea is flat and sandy, while the eastern coast is rocky with many islands.

Aegean Sea　*noun*
The Aegean Sea is a **gulf** of the **Mediterranean Sea**. It lies between Greece and Turkey. The Aegean Sea contains many islands.
They sailed their yacht on the Aegean Sea.

Agulhas Current　*noun*
The Agulhas Current is an **ocean current** that flows along the south-east **coast** of Africa.
The Agulhas Current carries warm water south towards the tip of Africa.

albatross　*noun*
An albatross is a large **seabird** with very long, narrow wings which help it glide just above the ocean waves. Albatrosses snatch fish from the surface of the sea as they fly.
An albatross spends almost its whole life at sea.

algae (singular **alga**)　*plural noun*
Algae are a group of simple, plant-like creatures which make their own food by the process of **photosynthesis**. They range in size from **microscopic**, one-celled forms to large **kelps** over 100 metres long. Algae are usually brown, green or red.
Most algae can be found in water or damp places.

algin　*noun*
Algin is a chemical made from **brown seaweeds**. It is used to set ice cream and puddings, to finish paper, to apply dyes and inks, and in medicines.
In California, in the United States of America, large quantities of seaweed are harvested in order to extract algin.

amphibious ship　*noun*
An amphibious ship is a warship that lands troops, weapons and vehicles on a **beach** during an amphibious assault. Some amphibious ships unload onto the beach itself. Others transfer troops and cargo into helicopters, landing craft, or small amphibious tractors called amtracs.
The troops jumped out of the amphibious ship onto the beach.

amphipod *noun*
An amphipod is a small, shrimp-like animal with a flattened body. **Sandhoppers** and beach fleas are kinds of amphipod. Most amphipods are **scavengers**.
Many amphipods live on the sea-bed.

anchor *noun*
An anchor is a heavy iron weight with two hooks at the end and a ring to which a cable is attached. The other end of the cable is fixed to a ship's deck. The hooks dig into the sea-bed. The anchor is used to hold the ship securely in a particular place.
An anchor must be very heavy, or the ship will drag it along the sea-bed.

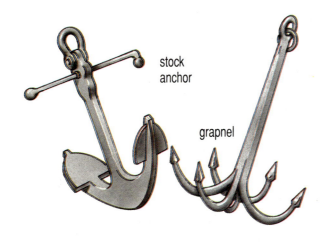

stock anchor

grapnel

anchovy *noun*
An anchovy is a small, slender, silvery **fish**. An anchovy's lower jaw tucks inside its upper jaw, so it looks as if it has no chin. Anchovies travel in large **shoals** near the coast and feed on **plankton**. They are caught in large numbers for food and bait.
Anchovettas are also members of the anchovy family.

anemone *noun*
An anemone is a cup-shaped animal. It has a ring of **tentacles** around its mouth which is used to catch small water animals. Stinging cells on the tentacles inject poison into the **prey** to paralyse it.
Anemones spend most of their life fixed to rocks or seaweeds.

anemonefish ► **clownfish**

anglerfish *noun*
An anglerfish lives on the sea-bed. It has an enormous mouth. Part of the dorsal fin forms a 'fishing rod', or lure, which has a fleshy tip. This dangles above the fish's mouth. As the rod wiggles, small fish think it is a worm and come to eat it. Then the anglerfish opens its huge mouth and sucks them in.
Anglerfish have colours that match the sea-bed.

Antarctic Convergence *noun*
The Antarctic Convergence is the area where warm water, blown towards the South Pole by the **Trade Winds**, meets cold water, blown towards the **Equator** by the **Westerlies**. This happens between **latitudes** 50 degrees and 60 degrees south.
At the Antarctic Convergence, cold water is forced down into the ocean depths.

Antarctic Ocean *noun*
The Antarctic Ocean surrounds Antarctica. It is made up of parts of the Atlantic, Pacific and Indian Oceans. These parts lie between the coast of Antarctica to the south and the **Antarctic Convergence** to the north.
Except in mid-summer, the Antarctic Ocean is covered in ice.

anticyclone ► **high pressure area**

aquaculture *noun*
Aquaculture is the farming of **fish**, **shellfish** and **seaweeds**. The fish are kept in small ponds or in cages suspended in **brackish** or salty water, and are given special food. The main aim of aquaculture is to produce food for people. Cultivated fish make up about 12 per cent of the total world fish consumption. In Asia, farmers cultivate large quantities of seaweed for food. The process of rearing fish and shellfish in cages in the **sea** or in **estuaries**, is called **mariculture**.
Oysters are reared by aquaculture in large estuaries.

aqualung *noun*
An aqualung is a device that allows a diver to breathe under water. It is made up of an aluminium tank, which can be filled with air up to a pressure of 3,000 pounds per square inch. A tube runs from a valve in the top of the tank to a demand valve that is held in the diver's mouth. When the diver breathes in, the demand valve supplies air. When the diver breathes out, the demand valve pushes air out into the sea water.
He put on his aqualung for the dive.

aquarium (plural **aquaria**) *noun*
An aquarium is a container used to house animals that live under water. Aquaria are usually made of glass. Most aquaria contain plants which provide oxygen for the animals.
They went to see the sharks in the aquarium.

Arabian Gulf *noun*
The Arabian Gulf is an arm of the **Arabian Sea** between the Arabian Peninsula and South-west Iran.
There are many oil fields around the shores of the Arabian Gulf.

Arabian Sea *noun*
The Arabian Sea is that part of the **Indian Ocean** lying between Saudi Arabia and India. The Indus and Narmada Rivers run into the Arabian Sea. In the past, the Arabian Sea was an important shipping route.
Oil tankers cross the Arabian Sea on their way from Saudi Arabia to India.

archerfish *noun*
An archerfish is a small fish that lives along the coasts of South-east Asia and Australia. It shoots a stream of water droplets through its mouth to knock down insects that have settled nearby or are flying above the water. Then it eats them.
The archerfish can shoot down insects over one metre away.

archipelago *noun*
An archipelago is a group of islands.
The islands of Indonesia form the largest archipelago in the world.

Arctic Ocean *noun*
The Arctic Ocean is the world's smallest ocean. It surrounds the North Pole. It is almost completely enclosed by the **coasts** of Alaska, Canada, Greenland, Norway and Russia.
Most of the Arctic Ocean is covered by ice all year round.

arthropod *noun*
An arthropod is an animal that has a body covered in a hard outer shell. The body is divided into sections called segments. All arthropods have jointed legs. As an arthropod grows, it sheds its hard outer shell and grows a new, larger one. This process is called moulting. Insects, spiders, **crustaceans**, centipedes and millipedes are all kinds of arthropod.
Young arthropods go through many changes as they grow up.

barnacle
horseshoe crab
crab
brine shrimp

arrow worm *noun*

An arrow worm is one of the most common animals found in **plankton**. It is a small, worm-like animal that can be between three millimetres and 100 millimetres long. An arrow worm has a transparent, arrow-shaped body. Arrow worms have narrow fins on each side of their body, and a tail fin. They feed on the **microscopic** animals in plankton.

The arrow worm ate the newly-hatched herring.

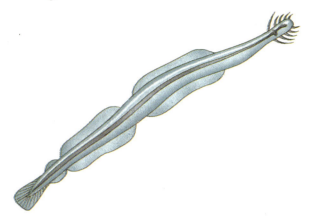

astrolabe *noun*

An astrolabe was an instrument used by early astronomers and navigators. It was invented by the Greek Hipparchus in 150 BC. The astrolabe was used for plotting the height and position of the Sun, Moon or stars. The astrolabe has been replaced mainly by the **sextant**, which is more accurate.

The sailor checked the position of the Sun by looking at the astrolabe.

Atlantic Ocean *noun*

The Atlantic Ocean is the second largest **ocean** in the world. It lies between North America, South America, Europe, Africa and Antarctica and covers about 106 million square kilometres, including gulfs and bays. One-fifth of the Earth's surface is covered by the Atlantic Ocean. The Atlantic Ocean is about 4,000 metres deep.

Ascension Island can be found in the Atlantic Ocean.

atmospheric pressure *noun*

Atmospheric pressure is pressure on the Earth's surface caused by the weight of the air above.

The effect of atmospheric pressure is like having a kilogram weight pressing down on every square centimetre of your body.

atoll *noun*

An atoll is a ring-shaped **coral** island or group of islands, which surrounds a **lagoon**. An atoll forms on a submerged mountain. The mountain once rose above the waves to form an **island**. A **coral reef** then grew up around its coast. Later, the sea-floor either sank or the **sea-level** rose but the reef kept growing upwards. Now, it is all that is visible of the island.

The Maldive Islands in the Indian Ocean are a group of atolls.

auk *noun*

An auk is a **seabird**. Auks have short legs that are set far back on the body. This makes it easier for them to swim, but gives them a clumsy, waddling walk on land. There are 22 species of auk. The auk family includes many birds such as guillemots, razorbills, puffins, murres and murrelets. They live in the North Pacific, North Atlantic and Arctic Oceans. Auks spend most of their time at sea, where they catch **fish** and **shellfish**.

Auks nest in huge colonies on cliffs.

puffin

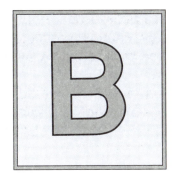

bacteria (singular **bacterium**)   *plural noun*
Bacteria are tiny living things that are too small to see without a microscope. They are so small that a row of 1,000 bacteria can easily fit on a pinhead. Some bacteria make their own food by **photosynthesis**, others feed on decaying plants and animals. Bacteria multiply quickly, by dividing in two.
A single bacterium can produce 4,000 million million million bacteria in 24 hours.

bacterioplankton   *noun*
Bacterioplankton are **bacteria** that live in **plankton**.
The smallest animals in plankton feed on the bacterioplankton.

baleen   *noun*
Baleen, or whalebone, is a horny material that grows like a fringe from the upper jaws of some kinds of **whale**. A single piece of baleen is about 0.6 centimetres thick. In larger species of whales, there can be up to 300 pieces of baleen, some of which can measure up to three metres long. Baleen is used to sift **plankton** from the water. The whale gulps in a huge mouthful of water, squeezes it out through the baleen and then swallows the plankton.
The whale licked up the plankton that was trapped on its baleen plates.

ballast   *noun*
Ballast is a heavy material such as gravel or **sand**. It is placed in a ship's hold to make the ship more stable and prevent it from capsizing in heavy seas.
They loaded the cargo ship with ballast.

Baltic Sea   *noun*
The Baltic Sea is a **gulf** of the **Atlantic Ocean**. It is enclosed by the **coasts** of Denmark, Sweden, Finland, Russia, Estonia, Latvia, Lithuania, Poland and Germany. Many large rivers flow into the Baltic Sea, so the water is not very salty.
In winter, the northern part of the Baltic Sea freezes over.

barbel   *noun*
A barbel is a short, fleshy **tentacle** that is found near the mouth of some fish. Barbels are used for tasting and touching.
Some fish use barbels to help them find food on the sea-bed.

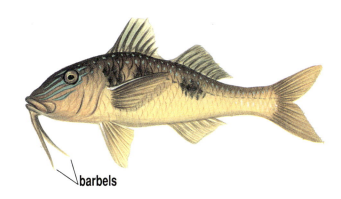

barbels

Barents Sea   *noun*
The Barents Sea is part of the **Atlantic Ocean**. It lies between the Russian islands of Novaya Zemlya and Franz Josef Land, the Norwegian islands of Svalbard, and the **coasts** of Norway, Finland and Russia.
The Barents Sea is a rich fishing ground.

barnacle   *noun*
A barnacle is a small **crustacean** which fastens itself onto objects under water. When feeding, the barnacle opens its shell and combs the water with its long, feathery legs. Some barnacles live attached to rocks, ships and the shells of other animals. Goose barnacles are attached to floating objects by stalks.
Young barnacles are not fixed to rocks, but swim in plankton.

barracuda *noun*

A barracuda is a fierce, slender fish with a long lower jaw and many large, sharp teeth. There are 21 **species** of barracuda in the world. The largest barracudas grow to three metres long, but most of them are smaller than this.
Barracudas strike at any shiny objects that move.

barrage *noun*

A barrage is a dam built across the mouth of an **estuary**. The energy of the tide rushing into the estuary is used to generate electricity. The water is trapped behind the dam until the tide has gone out. Then it is released to generate more electricity. This is called tidal power.
The first barrage was built across the Rance River in Brittany, France, in 1967.

barrier island *noun*

A barrier island is a long ridge of **sand** that forms a low **island** along, or parallel to, the **coast**. Barrier islands are formed when sand, gravel and shingle are deposited by the tide and waves.
Barrier islands protect the coast against storms that blow in from the sea.

barrier reef *noun*

A barrier reef is a long **coral reef** that runs along, or lies parallel to, the **coast**. It is separated from the shore by a large **lagoon**. Barrier reefs usually lie some distance from the shore.
The world's largest barrier reef is the Great Barrier Reef which is off the east coast of Australia.

Barrier Reef ▶ Great Barrier Reef

basalt *noun*

Basalt is a hard, dark rock which looks as if it is made up of very tiny pieces of rock. Basalt is formed when molten lava cools on the surface of the Earth, or on the sea-bed.
New ocean floor made of basalt is formed along the mid-oceanic ridges.

bathymetry *noun*

Bathymetry is the method used to measure the **depth** of the water in the ocean.
Bathymetry is used to work out the shape of the ocean floor.

bathyscaphe *noun*

A bathyscaphe is a manned vehicle used to explore under the sea. It is made up of a hollow steel sphere, which is designed to stand up to the huge pressure that is found at great depths.
Bathyscaphes are steered by electric propellers.

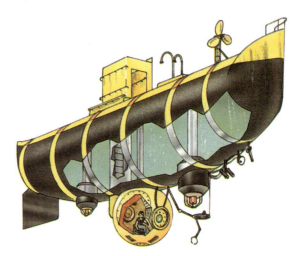

bathysphere *noun*

A bathysphere is a spherical container, which can be suspended underneath a ship to a **depth** of about 900 kilometres. It was invented in 1930 by William Bebe, so that undersea life could be studied.
Bathyspheres have now been replaced by the safer **submersible**.
The oceanographers went below the surface of the sea in the bathysphere to study the deep sea fish.

bathythermograph   *noun*
A bathythermograph is an instrument which is used to measure the temperature of the water at various **depths** in the ocean.
Some bathythermographs are towed behind ships and some are dropped from aircraft into the sea.

bay   *noun*
A bay is an area of sea which is enclosed by a curved section of the coast. Bays vary in size from a few hundred metres to thousands of kilometres across. The huge **Bay of Bengal** has an area of more than two million kilometres.
She could see the lights of the houses on the other side of the bay.

Bay of Bengal   *noun*
The Bay of Bengal is a large bay in the north-east **Indian Ocean**. It extends from Sri Lanka and India, to Bangladesh, Myanmar in Burma and the Malay Peninsula.
The Ganges, Brahmaputra and Irrawaddy Rivers flow into the Bay of Bengal.

Bay of Fundy   *noun*
The Bay of Fundy is an extension of the **Atlantic Ocean**. It lies between the Canadian provinces of New Brunswick and Nova Scotia. The Bay of Fundy has the largest rise and fall of any **tide** in the world. **Spring tides** here rise 21 metres.
The Bay of Fundy is famous for its fast-running tides.

beach   *noun*
A beach is a strip of **sand**, **gravel** or pebbles. It covers the seashore between the high and low water level of the **tides**.
A beach is formed by the action of **waves** breaking off small pieces of cliffs or soil at the land's edge and then grinding them down. Beaches are often used for recreation. Popular beach resorts lie along the Riviera on the Mediterranean coasts of southern France and northern Italy.
They searched the beach for seashells.

beach flea ► amphipod

beachcombing   *noun*
Beachcombing describes searching a beach to see what can be found on it. Shells, gemstones, **fossils**, pieces of glass and bottles, fish egg cases and the shells of **sea urchins** and **crabs** can all be washed up by the tide.
He had a collection of treasures found while beachcombing.
beachcomb   *verb*

Beaufort Scale   *noun*
The Beaufort Scale is a scale that is used to measure the force of the **wind**. The scale goes from calm, force 0, to hurricane, force 12.
A storm measures force 10 on the Beaufort Scale.

Beaufort Sea   *noun*
The Beaufort Sea is part of the **Arctic Ocean**. It is found off the northern **coast** of North America. The Beaufort Sea lies between the **Chukchi Sea** and Banks Island, off the Canadian coast.
The Beaufort Sea is often covered with floating ice.

beluga   *noun*
A beluga is a type of **whale**. It is also known as the white whale. The beluga is about four metres long, with a creamy white skin and a very round forehead. Belugas feed mainly on **fish**, **squid** and **crustaceans**. They are strong, slow swimmers.
The beluga lives in the Arctic Ocean and nearby waters.

bends   *noun*
The bends is a sickness divers get if they return to the surface of the water too quickly. Nitrogen and other gases bubble out of the blood and block the blood vessels.
The bends can cause permanent illness, or even death. The bends is treated by putting the diver in a **decompression chamber**.
A diver suffering from the bends has painful joints and finds it hard to breathe.

Benguela Current   *noun*
The Benguela Current is an **ocean current**, found off the **coast** of south-west Africa.
The Benguela Current carries cold water north towards the Equator.

benthic   *adjective*
Benthic describes the area at the bottom of the sea. It also describes the creatures living there. On the **continental shelf** and near the seashore, the benthic zone is quite shallow. In the deep sea, the benthic zone may be found at depths of more than 1,000 metres.
The opposite of benthic is pelagic.

benthos   *noun*
Benthos is the name given to the creatures living on the sea-bed, or in the **benthic** zone. The benthos includes **seaweeds**, **crabs**, **starfish**, **sea urchins** and **sponges**.
Many members of the benthos feed on the remains of dead plants and animals.

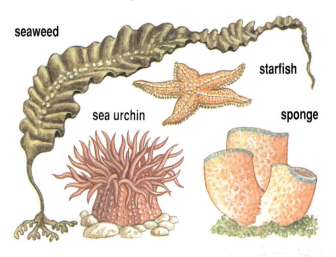

seaweed

starfish

sea urchin

sponge

Bering Sea   *noun*
The Bering Sea is part of the north **Pacific Ocean**. It is found between Russia, Alaska and the Aleutian Islands.
The Bering Sea is partly covered by ice in winter.

Bering Strait   *noun*
The Bering Strait is a narrow, shallow channel, found between Alaska and Russia. It connects the **Bering Sea** with the **Arctic Ocean**. During the Ice Age, when the sea-level was lower, the area now known as the Bering Strait was a strip of land. At that time, humans and animals could walk across it from Asia to North America.
The narrowest part of the Bering Strait is only 19 nautical miles wide.

Bermuda Triangle   *noun*
The Bermuda Triangle is an area of the **Atlantic Ocean**. It lies between Bermuda, Florida and Puerto Rico. The Bermuda Triangle covers 1,140,000 square kilometres. Many ships and aircraft have mysteriously disappeared without trace here.
There are many powerful currents in the Bermuda Triangle.

bioluminescence   *noun*
Bioluminescence is the light given off by animals and plants. Many deep sea animals produce light in a special part of their body called a **light organ**. Patterns of light organs on the bodies of **deep sea fish**, **shellfish** and **cephalopods** may help them find **prey** or mates in the dark waters of the deep sea.
Flashlight fish use bioluminescence to find their prey at night.
bioluminescent   *adjective*

biomass   *noun*
Biomass is the total weight, or mass, of all the living things in a particular area.
The biomass of a coral reef is much greater than the biomass of the deep sea floor.

bivalve ► page 13

bivalve *noun*

A bivalve is a **mollusc** that has two shells joined together by a hinge. When the shells are open, water is let in and out of the body cavity by two tubes called siphons. The body organs are enclosed in a soft membrane called the mantel. This develops into the shell. Bivalves come in many shapes and sizes.

Bivalves such as mussels, oysters, scallops and cockles are good to eat.

oyster

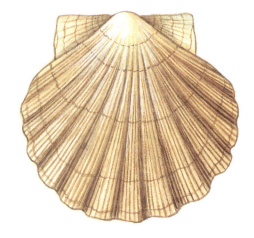

queen scallop

dog cockle

common mussel

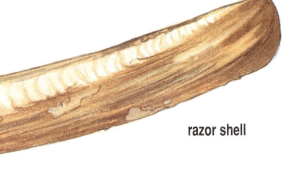

razor shell

Black Sea *noun*
The Black Sea is an inland sea in south-east Europe. It is surrounded by the Ukraine and Russia to the north, Georgia to the east, Turkey to the south, and Bulgaria and Romania to the west. The Black Sea covers approximately 453,000 square kilometres. Its deepest bed is 2,206 metres below the surface. The **Bosporus** links the Black Sea to the **Sea of Marmara**. Another narrow waterway connects it with the **Sea of Azov** in the north.
The River Danube flows into the Black Sea.

blenny *noun*
A blenny is a small fish. It has a blunt nose and long fins running along its back and along its belly. Many blennies have small tentacles, or **barbels**, on their head. They are found in **rock pools**, oyster beds, **mudflats** and among seagrasses.
Blennies have a smooth body, with no scales.

blowhole *noun*
A blowhole is a hole found in rocks. It is formed when the roof of a sea cave falls in. Plumes of spray shoot through the blowhole from the **waves** below, making a booming sound. Blowhole is also the name given to nostrils found on **cetaceans**. Powerful muscles and valves open the blowhole so the cetacean can breathe. Then, the blowhole snaps tightly shut.
The children were soaked when a jet of spray burst through the nearby blowhole as they stood on the cliffs.

blubber *noun*
Blubber is the name given to the fatty layer beneath the skin of **cetaceans**, **sirenians**, **pinnipeds**, and other sea mammals. Blubber helps to keep the animal warm and also helps it to float. **Whales** are still hunted for oil called ambergris, which can be extracted from their blubber.
The walrus's blubber wobbled as the huge animal hauled itself onto the rocks.

blue-green algae *plural noun*
Blue-green algae, or cyanobacteria, are a type of **bacteria**. They make their own food by **photosynthesis** just as algae do. They are very common in **plankton**, and also live on the moist surfaces of rocks and trees. In polluted water they may multiply rapidly to produce huge populations called blooms, which colour the water. Some kinds of blue-green algae release poisons into the water.
A bloom of blue-green algae had poisoned the shellfish so they were unfit to eat.

bony fish *noun*
Bony fish is the name given to fish which have bone in some part of their skeleton. Bony fish include most familiar fish, except for **sharks**, **rays**, **hagfish** and **lampreys**. They range in size from pygmy gobies, which are only 12 millimetres long, to marlin which are 4.5 metres long and ocean sunfish weighing over 900 kilogrammes. Bony fish are found all over the world in fresh water, salt water and hot springs.
There are over 20,000 different kinds of bony fish.

booby *noun*
A booby is a large **seabird** which lives near warm seas. It has long wings, a large bill, webbed feet and a long, wedge-shaped tail. Boobies are **diving birds**. They catch prey by plunge-diving into the sea. Boobies eat mainly **flying fish** and **squid**. They live in flocks on remote island and coastal cliffs.
The booby dived into the water to catch the fish.

bore *noun*

A bore is a giant flood **wave** that travels up a river against the usual flow of the water. Bores form in funnel-shaped **estuaries** when a very high tide pushes a huge amount of sea water into the estuary. The bore on the River Qiantong Jiang in eastern China reaches a height of 7.5 metres and a speed of 27 kilometres an hour.
They could hear the bore coming when it was still 22 kilometres away.

Bosporus *noun*

The Bosporus is a narrow waterway or **strait**, connecting the **Black Sea** with the **Sea of Marmara**. It is 31 kilometres long.
She gazed across the Bosporus at the mosques of Istanbul.

bow *noun*

The bow is the front part of a ship. It begins at the point where the sides of the ship start to curve inwards.
He stood in the bow of the ship watching the shore getting nearer and nearer.

boxfish *noun*

A boxfish has a hard-covered, box-like body. It is so stiff that only the eyes, mouth, fins and tail can move, so it cannot swim very well. If it is attacked, the boxfish gives off a poisonous substance which can kill nearby fish. Cowfish and trunkfish belong to the boxfish family. Boxfish are found in warm tropical seas all over the world.
Boxfish are covered with an armour of bony plates.

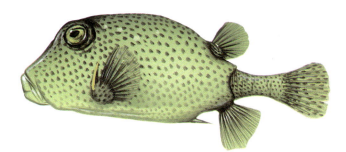

brackish water *noun*

Brackish water is salty, although not as salty as sea water. Brackish water contains from 0.5 to 17 parts of salt to every 1,000 parts of water.
Freshwater fish cannot survive in the brackish water of an estuary.

Brazil Current *noun*

The Brazil Current is an **ocean current** that flows down the east coast of South America.
The Brazil Current carries warm water from the Equator down to the South Atlantic Ocean.

breaker *noun*

A breaker is a **wave** that breaks on the **coast** or over a **coral reef**. Breakers form when waves reach shallow water. As the water drags against the sea-bed, it slows down. The water at the top of the wave overtakes the water at the bottom, so it topples forwards to form a breaker.
They enjoyed surfing in the huge breakers off the Pacific coast of America.

breakwater *noun*

A breakwater is a barrier protecting the shore from breaking waves.
The yacht sailed past the breakwater into the calm waters of the harbour.

brine *noun*

Brine is the name given to very salty water, especially to a concentrated solution of common salt, or sodium chloride.
He placed the meat in brine to preserve it.

brittlestar *noun*

A brittlestar is an **echinoderm**. It usually has five long, flexible arms which it uses for moving. Brittlestars scavenge for food on the sea-bed. The brittlestar has thin tubes called tube-feet on the underside of its arms. It uses them to breathe, to make small movements and to bring food to its mouth.
Dead fish that sink to the ocean floor are soon attacked by brittlestars.

brown seaweed *noun*

Brown seaweed contains a greenish-brown substance that takes in, or absorbs, light for **photosynthesis**. There are over 1,000 different kinds of brown seaweed. Some are whip-like, some are feathery and some have leathery cups. **Kelps** have ribbon-shaped fronds, or leaves, while wracks have branching fronds. Some brown seaweeds have air bladders to help them float up to the surface of the water. Brown seaweeds are collected commercially as a source of **fertilizer**, iodine, potash and **algin**.
Most brown seaweeds live in shallow water.

Bryde's whale *noun*

The Bryde's whale is a rorqual, or finback whale. This means that it has long grooves on its throat and chest, and a dorsal fin. Bryde's whales are a warm water species. They live in tropical and subtropical seas. Bryde's whales are bluish-grey in colour, with a white abdomen.
Bryde's whales can be up to 14 metres long.

buoy *noun*

A buoy is a floating marker in the sea or in a river. Lines of buoys are used to mark safe channels of deeper water in **estuaries** and near **reefs**. Buoys also mark areas of shallow water near the coast.
The sailors did not notice the buoys warning of shallow water, and the ship went aground.

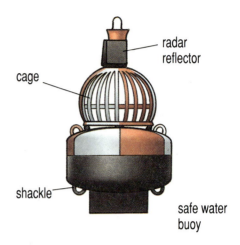

safe water buoy

buoyancy *noun*

Buoyancy is the way some objects tend to float in water. An object will float if it is lighter than water, otherwise it will sink. A piece of wood floats because there is a lot of air trapped in it. Many fish rise and sink in the water by varying the amount of air contained in their special **swim bladder**.
Without buoyancy the boat would sink.

butterflyfish *noun*

A butterflyfish is a small, brilliantly coloured fish which is found on **coral reefs**. It has a deep body that is very narrow, and a small mouth, with brush-like teeth. The French angelfish is an example of a butterfly fish.
Butterflyfish often feed on corals.

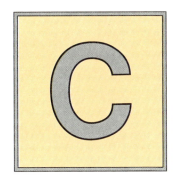

cable ► **submarine cables**

California Current   *noun*
The California Current is an **ocean current** off the west coast of the United States of America.
The California Current carries cold water from the North Pacific Ocean south, towards the Equator.

camouflage   *noun*
Camouflage is a kind of disguise which helps an animal to match its background and hide from its enemies.
The anglerfish's colours provide good camouflage against the sea-bed.

Cape Horn   *noun*
Cape Horn is the most southerly point of South America. It is found at the southern end of Horn Island, in Chile.
Cape Horn is famous for its stormy weather.

Cape of Good Hope   *noun*
The Cape of Good Hope is a **headland** in the far south-west of South Africa, near Cape Town.
The yacht sailed near the Cape of Good Hope.

capelin ► **smelt family**

carbon dioxide   *noun*
Carbon dioxide is a gas present in the air. It is released when animals breathe out and when fires burn. Plants use carbon dioxide in **photosynthesis**.
Carbon dioxide is a colourless gas which does not smell

carbonate   *noun*
A carbonate is a chemical that is formed when **carbon dioxide** reacts with other substances.
In the oceans, carbonates react with calcium compounds to form calcium carbonate.

cargo ship ► page 18

Caribbean Sea   *noun*
The Caribbean Sea is part of the **Atlantic Ocean**. It lies between the east **coast** of Central America, the north coast of South America and the West Indies.
Many ships travel through the Caribbean Sea on their way to the Panama Canal.

carnivore   *noun*
A carnivore is a flesh-eating animal.
Sharks are the ocean's fiercest carnivores.

cartilaginous fish ► page 20

Caspian Sea   *noun*
The Caspian Sea is an inland sea. It is surrounded by Russia and Kazakhstan in the north, and Azerbaijan, Iran and Turkmenistan in the south.
The Caspian Sea is the largest inland sea in the world.

catamaran   *noun*
A catamaran is a boat with two hulls. Catamarans were first used in Polynesia and Indonesia. They were powered by many men using oars to propel the boat along. In the 1900s, the design became more streamlined and more manageable.
Although it is hard to capsize a catamaran, once it is capsized, it is very difficult to turn upright again.

cell   *noun*
A cell is a **microscopic** building block found in the body of living organisms. A human is made up of billions of cells.
Cells make new living material and produce energy for body processes.

cargo ship *noun*
A cargo ship is a ship that is designed to carry goods from place to place. The area in the ship where the cargo is stored is called the hold. Cargo ships can carry general cargo or specialist cargo. Food carriers have freezer holds and car carriers have low decks to hold many vehicles.
The largest cargo ships are oil tankers which can be more than 1,300 metres long.

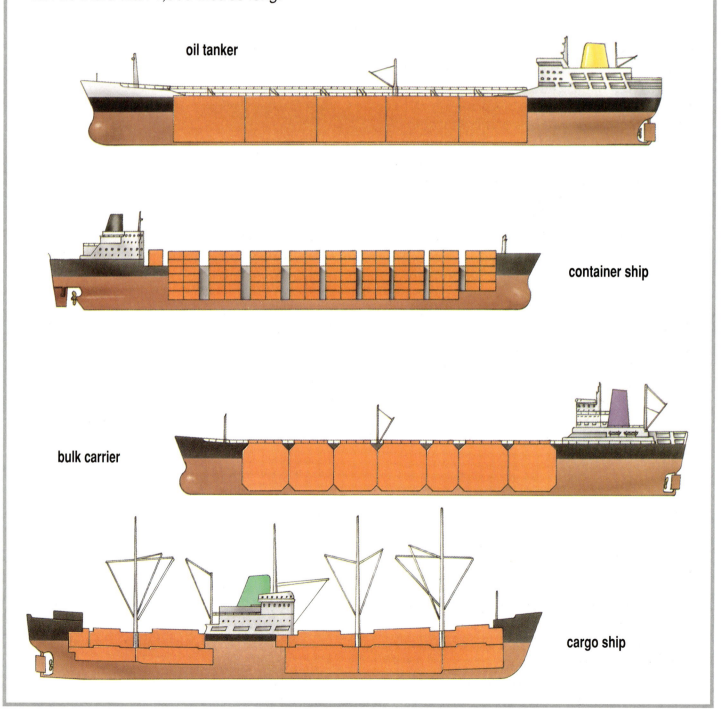

oil tanker

container ship

bulk carrier

cargo ship

18

cephalopod ▶ page 22

cetacean ▶ page 24

Challenger Deep   *noun*
The Challenger Deep is the lowest part of
the **Marianas Trench** in the western
Pacific Ocean. It is about 11,000 metres
deep.
*The Challenger Deep is the deepest point in
the ocean.*

chimaera   *noun*
A chimaera is a **cartilaginous fish** that lives
in very cold, deep water. It is sometimes
called a ghost shark or **rabbitfish**.
Chimaeras have a wedge-shaped or long,
narrow snout and tapering tail. Their eyes
are large for seeing in the dark and their gills
are covered by an **operculum**. Chimaeras
live in oceans throughout the world. They
feed on small fish and **molluscs**.
*Little is known about the chimaera's life
cycle.*

China Sea   *noun*
The China Sea is part of the western **Pacific
Ocean**. It is bordered in the north-east by
the Ryukyu Islands of Japan, and in the east
and south by Taiwan, the Philippines,
Borneo and Sumatra. The China Sea is
divided into the East China Sea and the
South China Sea, by the shallow Formosa
Strait between Taiwan and China.
*Fish from the China Sea supply the markets
of China and South-east Asia.*

chiton   *noun*
A chiton, or coat-of-mail shell, is a very
simple **mollusc** which is common in **rock
pools** and on underwater rocks. It has a
tough, protective shell, made up of eight
overlapping plates, which are bound
together by leathery flesh. Chitons use a
muscular foot to cling tightly to rocks. There
are more than 800 **species** of chiton.
*In the rock pool, chitons were grazing on
diatoms and small seaweeds.*

chronometer   *noun*
A chronometer is a very accurate clock.
The instruments inside a chronometer are
designed to react as little as possible to
movement and changes in temperature on
board a ship. Chronometers allow sailors to
calculate **longitude** accurately.
*The chronometer is held in place by gimbals,
or suspended rings, which allow it to stay flat
and level in relation to the horizon.*

Chukchi Sea   *noun*
The Chukchi Sea is part of the **Arctic Ocean**.
It lies between Russia's Wrangel Island in
the west, north-east Siberia and north-west
Alaska in the south, the Beaufort Sea in the
east, and the Arctic ice cap in the north.
*Whales, seals, walruses and polar bears live
along the coasts of the Chukchi Sea.*

cilia   *noun*
A cilia is an extremely small, hair-like
structure that is found on some
microscopic water animals. The larvae of
molluscs and **echinoderms** as well as
many microscopic creatures all have cilia.
*Bands of beating cilia propelled the larva
through the water.*

ciliate   *noun*
A ciliate is a **microscopic** animal. It is
covered in bands of **cilia** which it uses for
feeding and swimming. Most ciliates feed
on dead material, or **detritus**. A few are
parasites.
*Under the microscope they could see the
tiny ciliates.*

cartilaginous fish   *noun*

A cartilaginous fish is a fish with a skeleton made of cartilage. There are more than 500 **species** of cartilaginous fish. They include **sharks**, **rays** and **chimeras**. Their skeleton has the same layout as that of a bony fish, with a spine and pairs of **fins**. Cartilaginous fish do not have swim bladders. They live in all **oceans**.

The body of a cartilaginous fish is covered in rough scales.

There are about 350 species in the shark family. They are commonly found in warm seas. Sharks range in length from less than one metre to about 15 metres. They are all carnivores.

Shark skin is very rough and is covered in very small, tooth-like scales, called placoid scales, or denticles.

A shark has at least five rows of very sharp teeth. The front row is loosely attached and falls out after about 10 days, then it is replaced by the teeth which lie in the row behind.

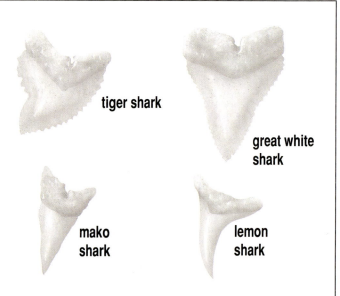

tiger shark

great white shark

mako shark

lemon shark

There are about 30 species of chimera, or ratfish. They are medium-sized fish with long eyes and a long, pointed tail. They live near the bottom of the ocean.

The Atlantic manta is the largest living ray. It weighs over 1,600 kilograms and its wing span can be more than six metres.

Rays swim by flapping their large fins to propel themselves forwards.

cephalopod *noun*

Cephalopod is the name given to a group of molluscs that includes, octopuses, cuttlefish, squid and nautiluses. The name cephalopod means 'head feet'. All cephalopods except nautiluses have an ink sac opening containing ink. This is released when they want to confuse enemies. or for camouflage. All cephalopods can change colour in less than a second.
Cephalopods are all carnivores.

There are about 50 different kinds of octopus living in warm, shallow seas. They range in size, from five centimetres to 9.6 metres long. The average lifespan of an octopus is three years. The main senses of the octopus are sight and touch.

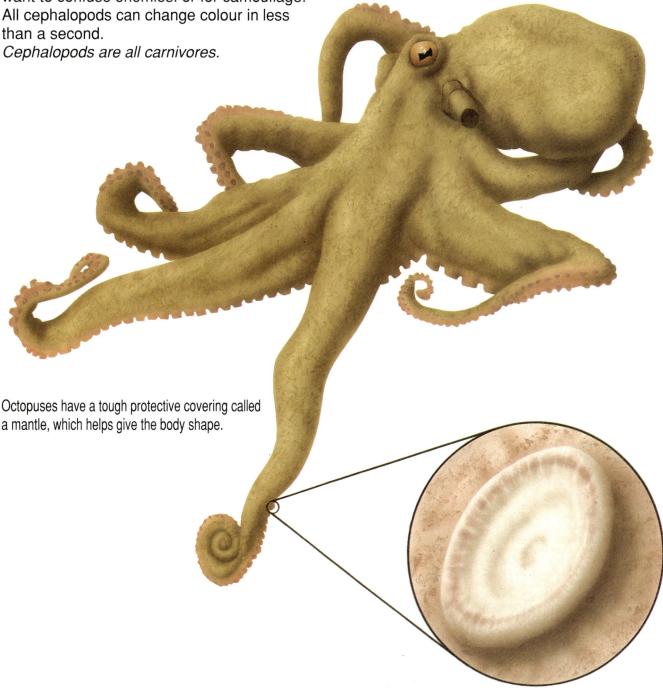

Octopuses have a tough protective covering called a mantle, which helps give the body shape.

The octopus has eight tentacles, with two rows of powerful suckers on each. These are used to grab objects, to feel things and to move around.

The chambered nautilus lives in a shell which is about 27 centimetres wide.

Cuttlefish range in size from eight centimetres to 1.8 metres in length.

Cuttlefish contain a spongy, chalky internal shell called a cuttlebone.

Squids live in the open sea and can swim very fast. They often swim in large shoals. Squids range in size, from less than 30 centimetres to almost 12 metres.

cetacean *noun*
A cetacean is a sea mammal. It belongs to a group of animals that includes **whales**, **dolphins** and **porpoises**. Cetaceans only live in the water, and breathe by means of a blowhole. This means that they must make frequent trips to the surface of the water for air. Cetaceans are protected by a thick layer of fat called blubber. This protects, or insulates, them from the cold.
Cetaceans have a streamlined body.

sperm whale

The sperm whale is the deepest diving sea mammal. It can swim down at least one kilometre and can hold its breath for more than an hour.

blue whale

The blue whale is the largest animal that has ever lived. It can be found in all oceans and can live up to 80 years.

24

bottle-nose dolphin

Dolphins live in schools. They are very
fast swimmers and can reach speeds of up to 40
kilometres per hour. There are 32 species in the dolphin
family. Dolphins live in all parts of the world.

common porpoise

There are six kinds of porpoise. They are usually smaller
than dolphins. The common porpoise is about 1.5 metres
long. Porpoises eat mainly fish and squid. They live in the
temperate seas of the northern hemisphere.

circulation ▶ **ocean circulation**

Circumpolar Current *noun*
The Circumpolar Current circles Antarctica from east to west. It is driven by the **Westerly Winds**. It is the only ocean current that extends from the surface of the sea all the way down to the **ocean floor**. It transports cold, deep water from the northern **Atlantic Ocean** into the **Pacific** and **Indian Oceans**.
The Circumpolar Current flows at a rate of about 200 million tonnes of water per second.

cleanerfish *noun*
A cleanerfish feeds on the **parasites** that live on other fish. Cleanerfish perform a kind of dance to advertise their service. They also have distinctive stripes so that other fish can recognize them. Fish may even open their mouth and raise their gill covers for the cleanerfish to go inside and feed on any parasites.
Several large fish were waiting among the corals in the reef, to be cleaned by the little cleanerfish.

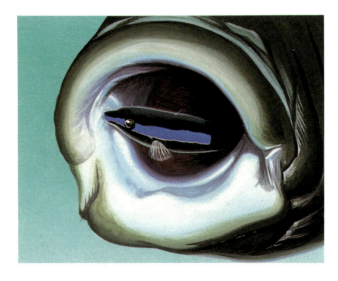

cleaning behaviour *noun*
Cleaning behaviour is the name given to the way certain kinds of fish and shrimps feed on the parasites that live on fish.
Cleaning behaviour is common.

cliff *noun*
A cliff is a steep rock face rising from flat ground, or from the side of a mountain. Cliffs are common at the **coast**, where **waves** have **eroded** the rocks.
Many seabirds nest on the steep cliffs of rocky islands.

clownfish *noun*
A clownfish, or anemonefish, is a brightly coloured fish that lives among the stinging **tentacles** of **sea anemones**. It feeds on scraps of food dropped by the anemone. By doing this, it helps to keep the anemone clean. This behaviour is a form of **symbiosis**. Scientists think that chemicals in the **mucus** covering the clownfish's body prevent the anemone from stinging it.
The anemone's stinging tentacles protected the clownfish from its enemies.

cnidarian ▶ page 27

coast *noun*
A coast is any place where the edge of the land meets the sea. Some coasts are gently sloping, with sandy **beaches**. Other coasts are wild and rocky, with steep **cliffs**. Coastal **habitats** include **estuaries**, coral **lagoons** and ice shelves.
Many ships have been wrecked on the dangerous, rocky coasts of Cornwall.

coastal water fish *noun*
Coastal water fish live along the shore in **rock pools** and in the waters above the **continental shelves**. These shallow waters are rich in nutrients, brought down by the rivers, so **plankton** grow well here. They attract **shoals** of **filter-feeding** fish like herrings, anchovies, pilchards and sardines. Cod and flatfish, such as plaice, skate and rays, feed on **invertebrates** found on the sea-bed.
Herring and mackerel are commercially important coastal water fish.

coat-of-mail shell ▶ **chiton**

cnidarian (plural **cnidaria**) *noun*
A cnidarian, or coelenterate, is a soft-bodied animal. It has a simple, cup-shaped body. At the top of the cup is a mouth, surrounded by stinging **tentacles**. **Anemones**, **corals** and **jellyfish** are all cnidarians. By squeezing water out of the top of the cup, or bell, jellyfish swim by **jet propulsion**.
Most cnidaria use their stinging tentacles to catch their prey.

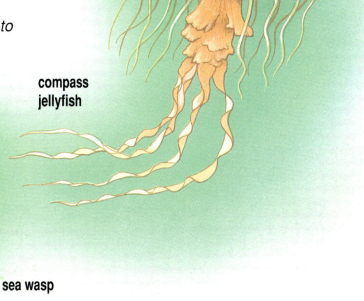

compass
jellyfish

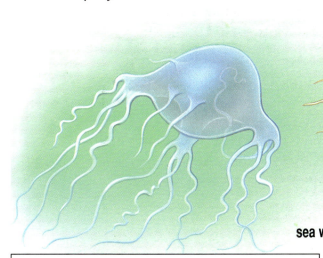

sea wasp

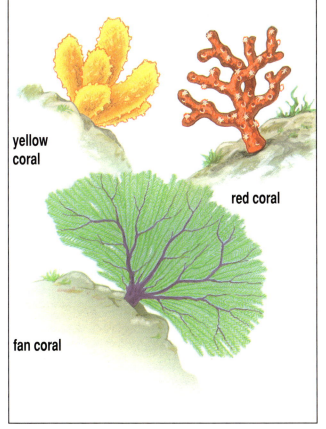

yellow
coral

red coral

fan coral

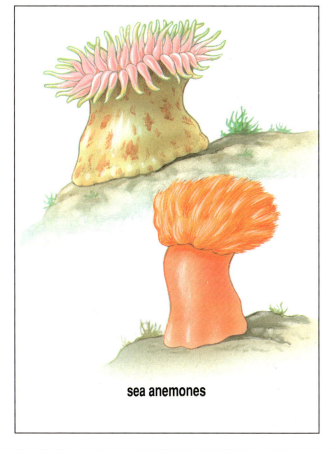

sea anemones

cod family   *noun*
The cod family contains many fish that are good to eat like cod, whiting, haddock and hake. They have large ventral fins in front of the pectoral fins. Many of them have fleshy **barbels** on their chin. Most members of the cod family feed on fish and **invertebrates** near the sea-bed.
Most members of the cod family live in large shoals near the sea-bed.

coelacanth   *noun*
A coelacanth is a strange fish. It has paired fins which have a stiff, fleshy base. The body of a coelacanth is covered in large, heavy scales. It has a wide, thick tail, with a middle section that can wiggle on its own. The coelacanth is the only survivor of an ancient group of fish over 400 million years old.
The coelacanth is found only in the extremely deep water between Africa and Madagascar.

coelenterate ► cnidarian

comb jelly   *noun*
A comb jelly, or sea gooseberry, looks like a jellyfish but it usually has only two **tentacles**. These are sticky, instead of stinging. Comb jellies swim by using rows of small comb plates. Each of these is fringed with **cilia**. Most comb jellies are egg-shaped, but the Venus' Girdle is a transparent, shimmering ribbon.
Comb jellies live in plankton.

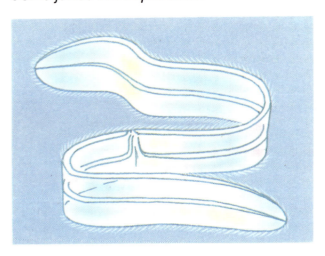

commercial fish ► page 30

compass   *noun*
A compass is an instrument which indicates the direction in which someone, or something, is travelling on the Earth's surface. A magnetic compass contains a magnet in the shape of a needle. The ends of the needle always point to magnetic north and south.
The sailor looked at his compass to see which direction his yacht was sailing in.

conch shell   *noun*
A conch shell is a **sea snail**, with a large, attractive shell.
Conch shells live on sandy sea-beds, in shallow tropical seas.

cone shell   *noun*
A cone shell is a tropical **sea snail**, with a cone-shaped shell. Some cone shells are so poisonous that their sting can cause death.
Many cone shells have striking markings on their shell.

conservation   *noun*
Conservation means taking good care of all the things around us. Conservation of living things means making sure we do not lose the great variety of plants and animals on the Earth. It involves protecting special **habitats** like **coral reefs**, **mangrove swamps**, seashores and coastal waters from **pollution** and destruction.
Nature conservation involves protecting animals and their habitats.

continental drift   *noun*
Continental drift describes how continents move over the surface of the Earth. The movement is very slow, only a few centimetres a year. Scientists believe that 200 million years ago, all the continents were joined together in one supercontinent called Pangaea. Then, Pangaea broke up and the pieces gradually moved to their present positions. Continental drift is part of the study of **plate tectonics**.
Scientists think that over millions of years continental drift pushed Africa away from South America.

continental shelf   *noun*
The continental shelf is part of the edge of a continent which continues under the sea as a shallow sea-floor. At the edge of the shelf, the sea-floor drops away very steeply. The width of the shelf varies from a few kilometres to about 400 kilometres.
The shallow waters above the continental shelf of western Europe are rich fishing grounds.

continental slope   *noun*
The continental slope is the steep slope at the outer edge of the **continental shelf**.
Many deep, underwater canyons have been carved in the continental slope.

convergence   *noun*
A convergence is the meeting of two **ocean currents**, flowing towards each other.
At a convergence, cold water usually sinks down below warmer water.

copepod   *noun*
A copepod is a tiny **crustacean** with two large antennae, or feelers, which it uses for swimming. A copepod is one of the most common animals in **plankton**. It lives in the deep oceans as well as in fresh water. Most copepods are between 0.5 and 2 millimetres long.
Copepods are important food for many kinds of fish.

coral   *noun*
A coral is a small, cup-shaped animal, or **polyp**. Its mouth is surrounded by stinging **tentacles**. Inside its body are microscopic **algae** which produce food by **photosynthesis** and then share it with the coral. This is why coral grows best in clear, sunlit water. Reef-building corals have skeletons of calcium carbonate, or limestone. They cannot live in water that is less than 18 degrees Celsius in temperature.
As new corals grow over the skeletons of dead ones, a rocky reef is built up.

coral bleaching   *noun*
Coral bleaching is what happens when **corals** die. They lose their colour and become white.
When water becomes cloudy with pollution, coral bleaching takes place and the reef dies.

coral island   *noun*
A coral island is a low, flat island formed from a **coral reef**. Coral islands are made up of mainly coral reef material. Coral reefs are made up of tiny sea organisms and their remains. They form and grow in warm, shallow water. Coral islands develop when the **sea-level** falls and the reef stands above the surface of the water. In time, the coral is eroded, to produce beaches of white coral **sand**.
Atolls may form coral islands that are ring-shaped.

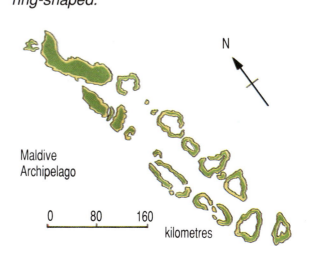

Maldive Archipelago

N

0   80   160
kilometres

commercial fish   *noun*

A commercial fish is a fish that is caught in order to be sold. About 65 million tonnes of fish are caught each year. **Herring**, **cod**, **mackerel** and **tuna** make up almost half the entire catch in the world. Most commercial fish are caught in the northern Pacific and northern Atlantic oceans.

Shrimps, crabs, lobsters and mussels are very important commercial fish.

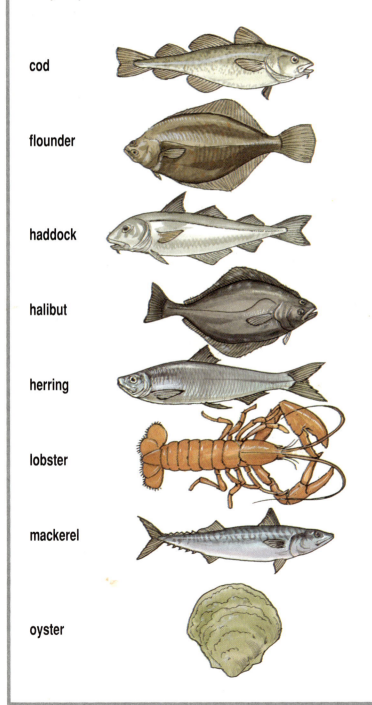

cod

flounder

haddock

halibut

herring

lobster

mackerel

oyster

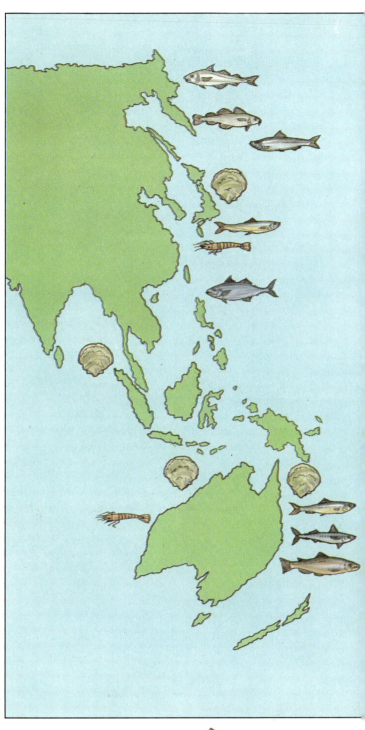

salm

sard

30

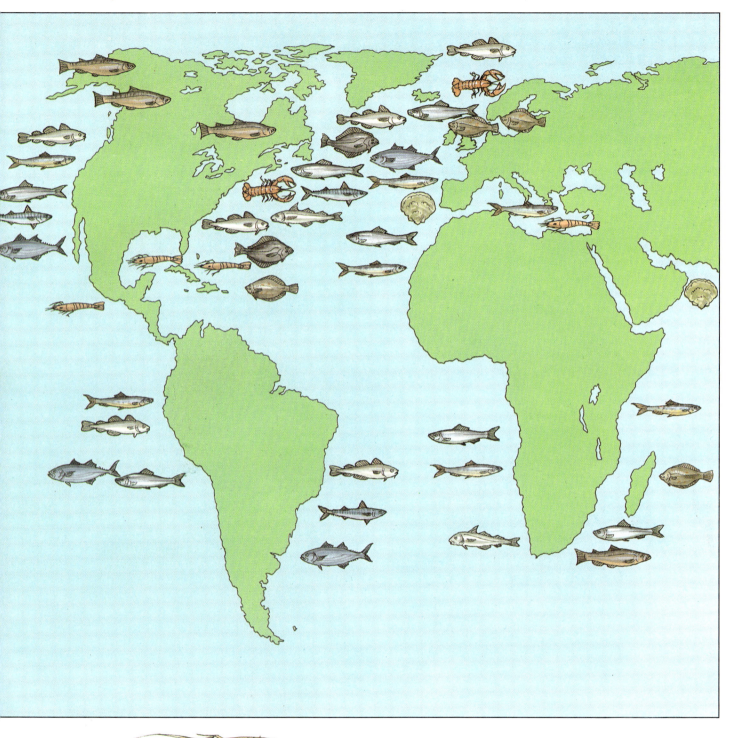

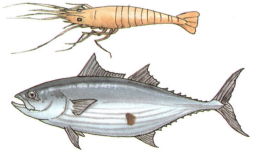

shrimp

tuna

coral reef fish   *noun*
A coral reef fish is brightly patterned, with stripes, spots and splodges of colour. This can be a kind of **camouflage**. The patterns break up the fish's shape when it is seen against the bright colours of **corals** and **sponges**. **Cleanerfish**, **surgeonfish** and **triggerfish** all use their colours to advertise or warn of danger.
Coral reef fish are some of the most colourful fish in the sea.

blue tang surgeon fish

royal gramma

clown triggerfish

long-nosed butterfly fish

lionfish

zebra pipefish

coral reef *noun*
A coral reef is a rock-like structure formed in clear, warm seas by colonies of small animals called **corals**. The limestone skeletons of corals build up over hundreds of years to make reefs. Different corals grow in different parts of the reef. On the outer slope, corals must be strong enough to cope with pounding by the waves. More delicate corals grow on the side of the reef that is nearest the land.
There are three kinds of coral reefs – fringing reefs, barrier reefs and atolls.

coral reef fish ► page 32

coralline alga (plural **algae**) *noun*
A coralline alga is a small **red seaweed**, covered in calcium carbonate. It is found on **coral reefs**. Coralline algae grow as crusts over the reefs surface, filling in cracks and cementing rubble together.
The skeletons of coralline algae help to bind the coral reef together.

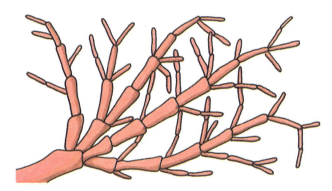

cordgrass *noun*
Cordgrass is a tall, stiff grass found in marshes and on **mudflats**.
Cordgrass covered the mudflats.

core *noun*
The core is the material found in the centre of the Earth. The inner part of the core is solid and is made up mainly of iron and nickel. The outer part is mainly composed of molten, or liquid, rock.
The core is about 6,940 kilometres in diameter.

Coriolis force *noun*
The Coriolis force is the force produced by the action of the Earth spinning on its axis. If an object is travelling from a pole in the direction of the **Equator**, it will curve to the right in the Northern Hemisphere and to the left in the Southern Hemisphere. It does not follow a straight line.
The Coriolis force makes the ocean currents travel in huge circles called gyres.

cormorant *noun*
A cormorant is a large seabird, with a long neck and short legs. Cormorants dive for fish, using their large, webbed feet to propel themselves underwater. They are mainly dark in colour, with glossy feathers. Some have white feathers underneath.
Cormorants live in large colonies on cliffs and rocks.

Cousteau, Jacques (1910–)
Jacques Cousteau is an ocean explorer and inventor. He invented the modern **aqualung** and designed underwater television cameras. He has also conducted scientific experiments to find out how people can work for long periods in deep water.
Jacques Cousteau has produced many books and films about the Earth's oceans.

cowrie *noun*
A cowrie is a **sea snail**, with a glossy, oval shell that tapers at one end. The opening of the shell is lined with notches on both sides.
Many cowrie shells have attractive colours and patterns, and are used to make jewellery.

crab *noun*
A crab is a **crustacean**. It is a small animal covered in a hard armour-like shell. The shield-like front part of a crab's body has a pair of eyes on short stalks, two short feelers or antennae, mouthparts and five pairs of legs. The front pair of legs is armed with pincers. The crab's short tail is folded under its body.
Some crabs feed on dead animals.

crustacean *noun*

A crustacean is an **invertebrate** animal, with many jointed legs. There are more than 22,000 **species** of crustacean, including **crabs**, **prawns** and **barnacles**. Most crustaceans live in the **sea**, but some live in fresh water. Crustaceans are **carnivores** and eat smaller crustaceans and **plankton**. Crustaceans have no bones. They have a hard external shell called an exoskeleton, which covers and protects the body. *Crustaceans can be between 1.3 centimetres and 25 centimetres long.*

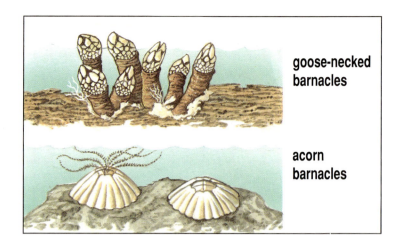

goose-necked barnacles

acorn barnacles

lobster

Claws and pincers for defence.

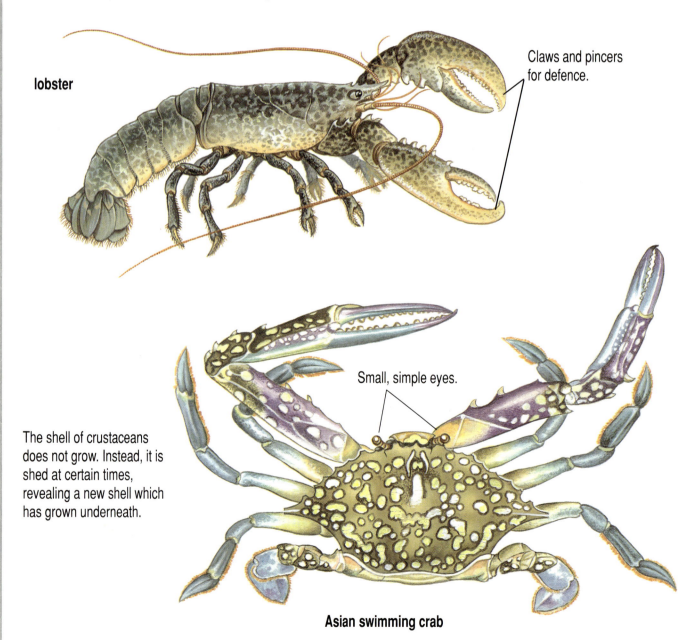

The shell of crustaceans does not grow. Instead, it is shed at certain times, revealing a new shell which has grown underneath.

Small, simple eyes.

Asian swimming crab

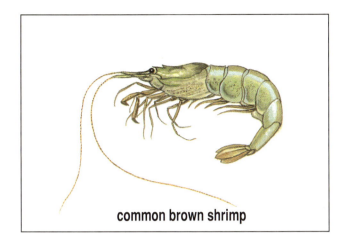

common brown shrimp

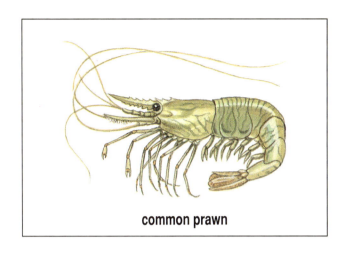

common prawn

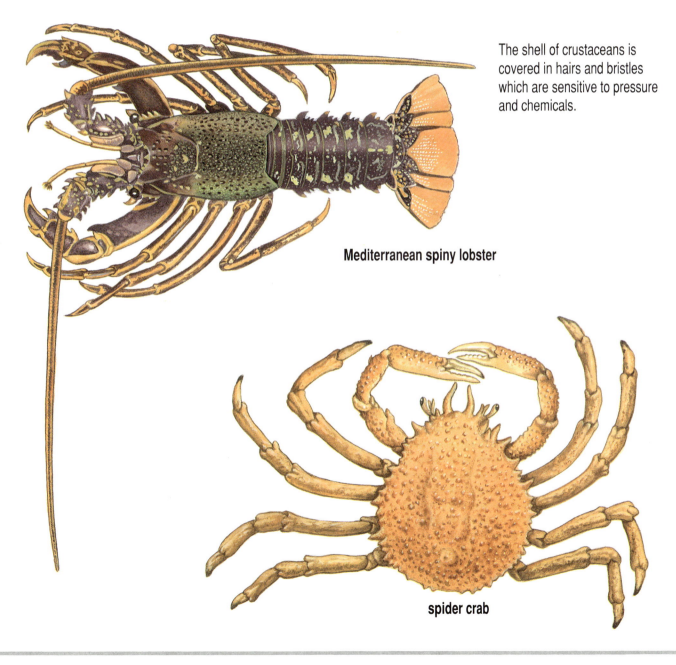

The shell of crustaceans is covered in hairs and bristles which are sensitive to pressure and chemicals.

Mediterranean spiny lobster

spider crab

crust *noun*
The crust is the outer layer of the Earth. It is mainly made up of granite and **basalt**. There are two types of crust. Continental crust is about 35 kilometres thick. Oceanic crust occurs under the oceans and is only about seven kilometres thick.
Under high mountain ranges, the continental crust may be up to 30 kilometres thick.

crustacean ▶ page 34

current ▶ **ocean current**

current meter *noun*
A current meter is a device used to measure currents under the sea. It is towed behind a ship, and turns like a weather vane to lie with the direction of the current. The current speed is measured by a turning propeller inside the **chronometer**.
He determined the strength of the current by looking at the current meter.

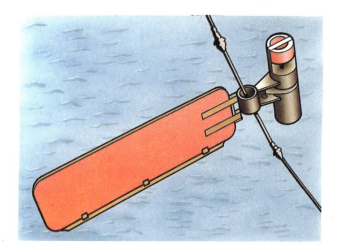

cuttlefish *noun*
A cuttlefish is a **cephalopod**. It has a flattened, torpedo-shaped body with a long narrow fin on each side. Cuttlefish have eight **tentacles** armed with suckers, and two longer tentacles with suckers at their tip. Inside their body is a hard shell, called a cuttlebone. Caged birds peck at cuttlebones to get extra calcium.
Cuttlefish use their long tentacles to capture small fish and crustaceans.

cyanobacteria ▶ **blue-green algae**

cyclone *noun*
A cyclone, or **depression**, is a system of **winds** circling around a centre of low **atmospheric pressure**. In the tropics, cyclones can be violent with very strong winds. They are sometimes called **hurricanes** or **typhoons**.
A tropical cyclone destroyed the banana plantations.

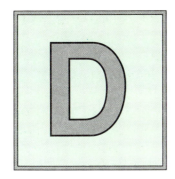

damselfish *noun*

The damselfish is a brightly coloured fish with a flattened body and spiny dorsal and anal fins. Damselfish live in shallow, warm seas. Common damselfish include zebrafish, garibaldis and clownfish.
Many of the most colourful coral reef fish belong to the damselfish family.

Dead Sea *noun*

The Dead Sea is a sea found between eastern Israel and western Jordan. It lies well below **sea-level**. Although it is fed by the River Jordan, no rivers drain it. Evaporation has left the water of the Dead Sea very salty.
The Dead Sea is so salty that a person can sit up in it and read a newspaper without sinking.

decompression chamber *noun*

A decompression chamber is a machine where divers go to recover from the **bends**. Here, they are put under high **atmospheric pressure**, which is very gradually lowered to normal atmospheric pressure.
He rose to the surface too fast, and had to spend several hours in a decompression chamber.

deep ocean current *noun*

A deep ocean current occurs when cold water sinks below warm water, or salty water sinks below fresh water.
For thousands of years, deep ocean currents have carried water from the Arctic to the Antarctic and then to the Pacific Ocean.

deep scattering layer *noun*

A deep scattering layer is a layer of the ocean which bounces back sound. This happens during **echo sounding**. Deep scattering layers are found at depths of about 300 to 400 metres by day, but rise close to the surface at night.
Scientists think deep scattering layers are due to fish such as hatchetfish and bristlemouths, that move to the surface to feed at night.

deep sea *noun*

The deep sea is a cold, dark place. In the upper part of the deep sea, fish can detect the silhouettes of other fish above them against the faint light coming from the surface. In the depths, there is no light at all. Here the temperature is around two degrees Celsius, so it is very cold. The weight of thousands of metres of water above creates a tremendous pressure.
In the deep sea, animals have to find their prey by smell, taste and touch.

deep sea fish ▶ page 38

delta *noun*

A delta is a wide fan of clay, sand, gravel and other sediment, put down, or deposited, by a river at the point where it enters a lake or sea. Deltas have very fertile soil, which makes them excellent agricultural areas.
A river usually splits into many channels as it crosses its delta.

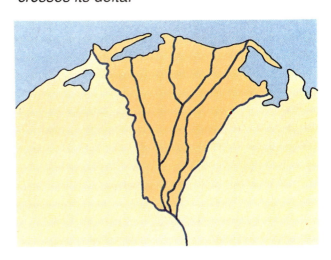

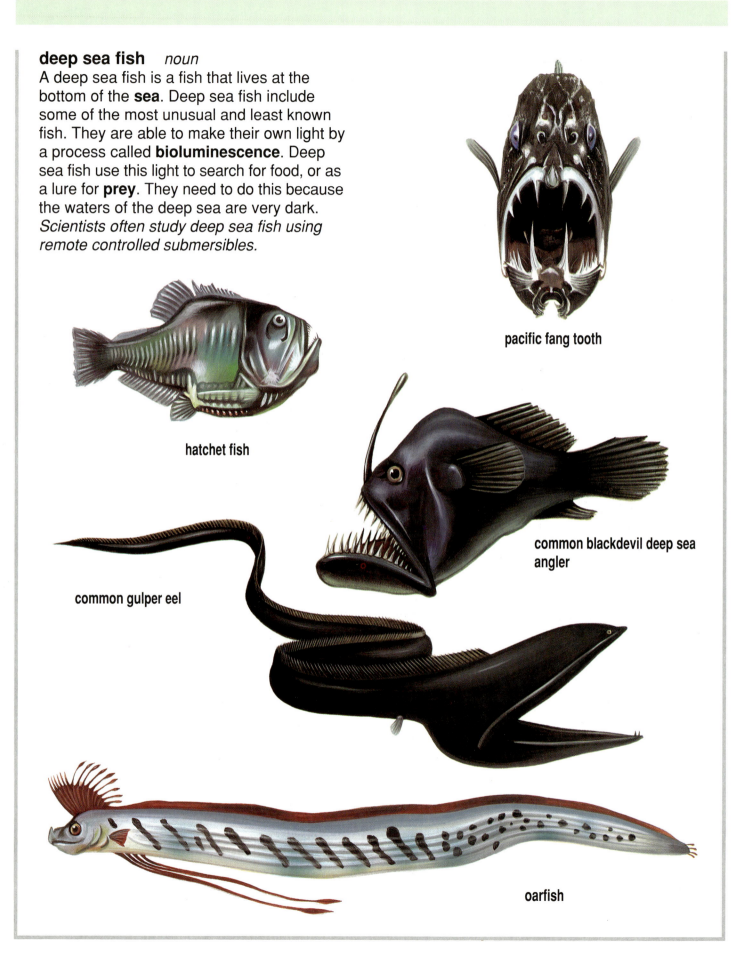

deep sea fish   *noun*

A deep sea fish is a fish that lives at the
bottom of the **sea**. Deep sea fish include
some of the most unusual and least known
fish. They are able to make their own light by
a process called **bioluminescence**. Deep
sea fish use this light to search for food, or as
a lure for **prey**. They need to do this because
the waters of the deep sea are very dark.
*Scientists often study deep sea fish using
remote controlled submersibles.*

pacific fang tooth

hatchet fish

common blackdevil deep sea
angler

common gulper eel

oarfish

demersal   *adjective*
Demersal describes anything that lives close to the bottom of the sea.
The goatfish is a demersal fish.

density   *noun*
Density is the weight, or mass, of a substance compared with its volume.
The density of water is one gram per cubic centimetre.

depression   *noun*
A depression is an area where the **atmospheric pressure** is low. Depressions bring cloudy skies, unsettled weather and often rain. They form over the ocean where cold air from the polar regions meets warm air from the tropics.
A depression was moving into the area from the Atlantic Ocean, bringing rain.

depth   *noun*
Depth is the measurement from the surface of the water downwards. It is usually measured in metres. Scientists use special words to describe the depth of the water and the depth of the **ocean floor**. Depth can also be described in terms of the amount of light present. The photic zone for example, receives enough light for algae to carry out **photosynthesis**. The disphotic zone is often called the twilight zone. In the aphotic zone there is no light at all.
At a depth of 1,000 metres, the ocean is cold and dark.

Depth	light zones	water zones	sea-floor zones
0–200 m	photic	epipelagic	sublittoral
200–1,000 m	disphotic	mesopelagic	bathyal
1,000–4,000 m	aphotic	bathypelagic	bathyal
4,000–6,000 m	aphotic	abyssalpelagic	abyssal
6,000–10,000 m	aphotic	hadalpelagic	hadal

desalination plant   *noun*
A desalination plant is a place where salt is removed from sea water. This process is called desalination. It makes water suitable for drinking.
Most of the drinking water in Saudi Arabia is produced by desalination plants.

detritivore   *noun*
A detritivore is an animal that feeds on the broken-up remains, or **detritus**, of dead animals, plants and **algae**.
The dead remains of marine animals are sucked up by detritivores, and made into living tissue again.

detritus   *noun*
Detritus is the broken-up remains of dead plants, animals and **algae**. Detritus is also the name given to loose, rocky material produced by the wearing away of rocks. This detritus collects on the sea-bed and at the foot of cliffs.
In the oceans, detritus sinks to the ocean floor, where it provides food for animals of the deep.

diatom   *noun*
A diatom is a microscopic **alga** that floats among **plankton**. Diatoms have transparent, glossy shells made up of two parts, which overlap like a pill box. Some diatoms live alone, but others are joined together in chains or plates.
The delicate patterns on the shells of diatoms have given them the name 'jewels of the sea'.

dinoflagellate   *noun*
A dinoflagellate is a **microscopic** shell alga that lives in **plankton**. Each dinoflagellate has two flagella. These are whip-like, beating hairs which are used for swimming. The flagella lie in grooves in the shell. Some dinoflagellates are poisonous and can kill fish.
Some dinoflagellates carry out photosynthesis, but others take in food particles from the sea water.

diurnal tide   *noun*
A diurnal tide is a movement of water. It is a **tide** that has only one high water and one low water period each day.
The tides along the coasts of China are diurnal tides.

diving   *noun*
Diving is the technique used by humans to descend below the surface of the sea. A free diver, or a diver with no equipment, can swim down to about nine metres, and stay submerged for as long as two minutes.
A diver with an **aqualung** needs to wear a thermal suit for warmth and a weight belt to give neutral **buoyancy**. Deep sea divers repair ships, conduct research and recover valuable objects.
For diving, a person needs a pair of flippers and a facemask.

diving bell   *noun*
A diving bell is a large, hollow container which is filled with air and is open at the bottom. It is used to carry divers and their tools to work on the bottom of the sea or river. A pipe at the top of the diving bell allows a supply of air to be pumped into the bell from the surface of the water. The first diving bell to be used was invented by the Italian Auglielmo del Lorena in 1531.
Modern diving bells are made of steel and are fitted with electric lights, telephones and other equipment.

diving bird   *noun*
A diving bird is a bird that hunts for fish or **shellfish**, or feeds on weeds under water. Gannets, **boobies**, brown **pelicans** and terns are high divers. They fold their wings and drop like arrows from the sky. Cormorants, **auks** and diving ducks paddle along the surface of the water and then dive. Most diving birds have webbed feet which give a large area for paddling and swimming.
Puffins are diving birds that have ridges on their bills for gripping slippery fish.

diving mammal   *noun*
A diving mammal is a mammal that dives under water, in search of food. Some reach depths of over 1,000 metres. Diving mammals have to cope with great pressure and low temperatures, and must store enough oxygen for the dive. Seals lower their heart rate during dives, so they need less oxygen. Their thick **blubber** helps keep them warm, and they have special arrangements of blood vessels to save heat.
Most diving mammals can close their ears, nostrils or blowholes to keep out water.

diving suit   *noun*
A deep sea diver uses a strong, shell-like diving suit with watertight joints. These suits are very strong and light and can be kept at surface pressure inside and allow a diver to descend to about 400 metres.
The diver wore his diving suit to keep warm in the cold water.

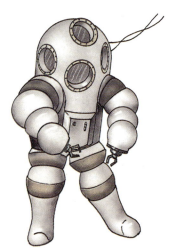

dogfish *noun*
A dogfish is a small **shark** that lives in shallow water near the sea-bed. Dogfish have short, stout teeth for feeding on molluscs, **crustaceans** and other fish.
Fishmongers sell dogfish as 'rock salmon' or 'huss'.

doldrums *noun*
The doldrums is an area of the ocean where the winds are often very light. The doldrums lie on either side of the **Equator**, where the north-east and south-east **Trade Winds** meet and warm air rises.
The sailing ship had hardly moved for days because of the calm weather of the doldrums.

dolphin *noun*
A dolphin is a small **cetacean**. Dolphins live in large **schools** and hunt fish, which they grip with their many pointed teeth. Dolphins use **echo-location** to find their way and communicate under water. There are 32 **species** of dolphin, including the orca, or **killer whale**.
Dolphins can reach speeds of up to 40 kilometres per hour.

dredging *verb*
Dredging is the process of clearing silt out from the bottom of rivers, harbours and other areas of water, so that ships can use them safely. The machines that do this are called dredgers. Dredgers are usually powered by diesel or steam engines.
The machines had started dredging in the port, so that ships can navigate through it easily.
dredge *verb*

drift net *noun*
A drift net is a long net which is strung out like a curtain across the ocean. Floats at the top and lead weights at the bottom keep it vertical. Fish that swim into drift nets are trapped by their **gill** covers when they try to back out. Drift nets trap any animal that swims into them, including **sea turtles**, **dolphins** and many fish of no commercial value. They are slowly being phased out.
Drift nets kill so many marine animals that they have been called 'walls of death'.

drilling *verb*
Drilling is the process of boring holes into rocks or into the sea-floor to look for oil or natural gas. A long, rotating pipe with sharp teeth cuts though the rocks. It is supported by a platform called a rig. If oil or gas is found, it flows up through the drill hole which becomes an oil well.
The oilmen pumped water and mud down the drill pipes to keep the bit cool while they drilled.

dugong ▶ **sirenian**

41

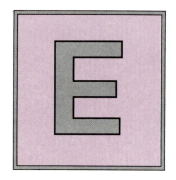

Earth   *noun*
The Earth is the planet on which we live. It is made up of several layers. The inside of the Earth is very hot. The centre of the Earth is a core of solid metal, surrounded by molten, or very hot, liquid metal. Then comes the mantle, which is made of molten rock. This is more than 70 kilometres from the Earth's surface. The outer layer of hard rock is the crust. The crust is divided into **tectonic plates**.
The structure of the Earth is very different from the structure of the other planets.

earthquake   *noun*
An earthquake is a shock wave produced by a large, sudden movement of rocks in the Earth's crust. The shock makes the ground shake. Earthquakes are usually caused by movements of **tectonic plates**.
During an earthquake, buildings may fall down, gas and water pipes may break and flooding may take place.

East Australia Current   *noun*
The East Australia Current is an **ocean current** flowing down the east **coast** of Australia.
The warm water of the East Australia Current bathes the corals of the Great Barrier Reef.

East Greenland Current   *noun*
The East Greenland Current is an **ocean current** flowing down the east **coast** of Greenland.
The East Greenland Current carries cold water from the Arctic Ocean into the Labrador Sea.

East Pacific Rise   *noun*
The East Pacific Rise is a large chain of underwater mountains in the south-east Pacific Ocean. It is a **mid-ocean ridge** at the boundary of three of the great **tectonic plates** that make up the Earth's crust.
Molten lava wells up along the East Pacific Rise creating new ocean floor.

ebb   *verb*
Ebb describes the movement of water away from the shore as the **tide** falls.
The opposite of ebb is flow.

ebb current   *noun*
An ebb current is a current that flows when the **tide** is going out.
The boat was dragged out to sea by the ebb current.

ebb tide   *noun*
The ebb tide is the movement of water away from the shore when the **tide** is falling.
On some beaches it is dangerous to swim in the ebb tide.

echinoderm   ▶ page 43

echo-location   *noun*
Echo-location is a way of finding out the shape and position of objects by measuring how sound bounces, or echoes, off them.
Dolphins use echo-location to find their way under water.

echo-sounding   *noun*
Echo-sounding is a way of finding the **depth** of the **ocean floor** or of locating solid objects under water, by using an echo-sounder. An echo-sounder sends out sound waves which bounce back off the sea-bed. The echo-sounder calculates the depth or distance, by measuring the time between sending out the sound waves and receiving back the echo.
Oceanographers map the ocean floor by echo-sounding.

echinoderm   *noun*

An echinoderm is a small marine **invertebrate**. Most echinoderms are covered in hard plates of calcium carbonate, often armed with knobs or spines. They move by means of hundreds of tiny feet, or small, bendy tubes with suckers at the tip. *Most echinoderms live on the sea floor and feed on other animals or the remains of dead animals and plants.*

starfish

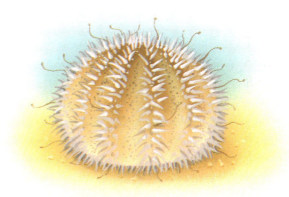

sea urchin

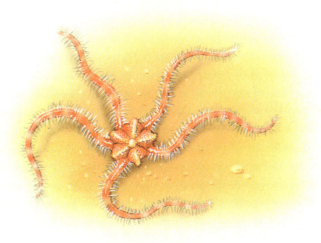

brittlestar

sand dollar

feather star

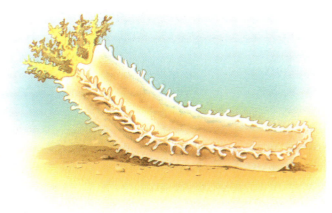

sea cucumber

ecosystem *noun*
An ecosystem describes a group of animals and plants and the environment, or habitat, in which they live. The seashore, a **rock pool** and the **ocean floor** are all types of ecosystem. The creatures that live in an ecosystem find most of their food and energy supplies within it. They are all part of the same **food web**.
Marine ecologists study marine ecosystems.

eddy *noun*
An eddy is air or water which flows in a different direction from that of the main **current**. Eddies form when water flows around an object or over an uneven surface. They also form when two masses of water flowing in different directions meet. Wind blowing over the water surface also causes eddies.
Many small eddies form where the Norway Current flows past the rugged Norwegian coast.

eel *noun*
An eel is a snake-like, **bony fish**. It has single, ribbon-like fins on its back and belly. Many eels have no scales.
When threatened by an enemy, eels can easily slide into crevices and hide.

eelgrass *noun*
Eelgrass is a grass-like flowering plant that grows on muddy **coasts** and **estuaries** between the high and low water mark. Eelgrass leaves can measure up to 1.8 metres long. They are often eaten by wild ducks and geese.
At low tide, a large flock of Brent geese flew down to graze on the eelgrass.

El Niño *noun*
El Niño is a climatic event that takes place about every 10 years. The **Trade Winds** in the tropical South **Pacific Ocean** become weaker than usual. This allows a warm current, called the El Niño Current, to flow down the western coast of South America instead of the usual cold current. Heavy rain then falls along the nearby coast.
Large numbers of fish and seabirds died during the last El Niño.

element *noun*
An element is a basic material. Everything in the universe is made up of elements. Elements cannot be split into different materials. About 90 different elements are found in nature.
Common salt contains the elements sodium and chlorine.

elements found in sea water

element	parts per thousand in sea water	element	parts per thousand in sea water
chlorine	19.5	carbon	0.028
sodium	10.77	nitrogen	0.0115
magnesium	1.29	strontium	0.008
sulphur	0.905	oxygen	0.006
calcium	0.412	boron	0.0044
potassium	0.380	silicon	0.002
bromine	0.067	fluorine	0.0013

endangered species *noun*
An endangered species is an animal or plant which is so rare that it is likely to disappear altogether. Most endangered species are protected by law.
The blue whale has been hunted so much that it is now an endangered species.

English Channel *noun*
The English Channel is a body of water between England and France. It connects the **Atlantic Ocean** and the **North Sea**.
Many cargo ships pass through the English Channel on their way to North Sea ports.

44

erode *verb*
To erode means to wear away the surface of something. The force of waves erodes the base of cliffs by breaking the rocks into smaller pieces.
There is a dangerous gap in the cliff path where the top of the cliff has been eroded.

Equator *noun*
The Equator is an imaginary line drawn around the Earth, half-way between the North and South Poles. It divides the Earth into two halves, the northern hemisphere and the southern hemisphere.
Countries lying on the Equator have warm, wet weather all year round.

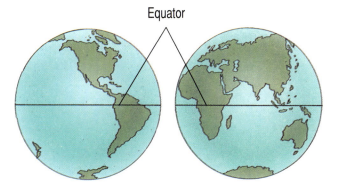

Equator

Equatorial Countercurrent *noun*
The Equatorial Countercurrent is an **ocean current** that flows eastwards, near the **Equator**. On each side of it, the **Trade Winds** cause **Equatorial Currents**, which push water westwards. The water accumulates in the west, and flows back towards the **doldrums** in the east as the Equatorial Countercurrent.
The ship was carried eastwards across the Pacific Ocean by the Equatorial Countercurrent.

Equatorial Current *noun*
An Equatorial Current is an **ocean current** that flows westwards near the **Equator**. The water of the Equatorial Current is driven by the **Trade Winds**.
Equatorial Currents bring warmth and moisture to the east coasts of Africa and South America.

estuary *noun*
An estuary is the mouth of a river where fresh river water meets the salty water of the sea. An estuary is affected by the **tides**. The water rises and falls as the tide ebbs and flows.
Animals and plants living in estuaries have to cope with large daily changes in the salt content of the water.

eutrophication *noun*
Eutrophication describes what takes place when lakes and seas receive too much **fertilizer**. This happens when sewage or industrial waste is released into the water. The extra nutrients help **algae** in the water to multiply very fast. Soon, there are so many algae that they can get no light. This means they cannot **photosynthesize**, so they die. The rotting algae use up the oxygen in the water, and this suffocates fish and other animals.
Algal blooms caused by eutrophication along the coast made the water unfit to swim in.

evaporation *noun*
Evaporation describes what happens when a liquid turns into a gas or vapour. Water turns into water vapour when it evaporates from the surface of the sea.
The inland sea shrank as the water level went down because of evaporation.
evaporate *verb*

Exclusive Economic Zone *noun*
An Exclusive Economic Zone, or EEZ, is the area of a country's **coast** which extends for 200 nautical miles out to sea. Under the United Nations Law of the Sea, individual states now have control over these waters. They are no longer common property.
The fishermen had to get licences to fish inside Iceland's Economic Exclusion Zone.

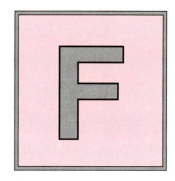

factory ship *noun*

A factory ship is the mother, or main ship, in a fleet of fishing or whaling boats. The other fishing boats bring their catch to the factory ship to be cooked, canned or frozen. This allows fishing fleets to spend long periods away from home.
Huge factory ships allow fishing fleets to travel far from home.

Falkland Current *noun*

The Falkland Current is a branch of the **West Wind Drift**. It flows up the east coast of Argentina in South America, then turns east to rejoin the West Wind Drift.
Icebergs from Antarctica are carried past the South American coast by the Falkland Current.

farming the sea ► mariculture

fathom *noun*

A fathom is an old English measure of length and **depth**. A fathom was the distance between a man's outstretched arms, from fingertip to fingertip.
One fathom measures about 1.83 metres.

fathometer *noun*

A fathometer is an instrument used on ships to measure the **depth** of water. A sound is sent down through the water. It is then echoed back from the bottom of the **sea**. The depth below the ship is calculated by measuring the time it takes for the sound to return.
The navigator used the fathometer to determine the depth of the water.

feather star *noun*

A feather star is an **echinoderm** with long, feathery arms attached to a central disc. Feather stars usually have five or ten arms. They fix themselves to rocks using small, jointed stalks. They then trap floating food **particles** on their sticky arms.
The feather star climbed up a coral and spread out its arms to catch some prey in order to feed.

ferry *noun*

A ferry is a kind of ship that carries passengers and vehicles across rivers, canals or narrow stretches of sea. Most ferries have a large opening at each end so they can be loaded and unloaded without being turned around.
They caught the ferry from Dover in England to Calais in France.

fertilizer *noun*

A fertilizer is a chemical or mixture of chemicals added to soil. Fertilizers make the soil rich, or fertile. As a result, plants or crops can grow in abundance.
Fishmeal makes a very good fertilizer.

filter feeder *noun*

A filter feeder is an animal that feeds by sieving, or filtering, small particles of food or tiny water creatures from the water. There are many different ways to filter food.
Filter feeders such as barnacles and shrimps use fringes of tiny bristles on their legs to sieve food from the water.

fin *noun*
A fin is a flattened structure attached to an animal's body. It is used for swimming and steering. The fins on a fish are supported by finger-like rays made of bone or cartilage. Fins may be arranged singly or in pairs. They are given special names according to their position on the body.
The boxfish has such a stiff body that it has to rely entirely on its fins in order to swim.

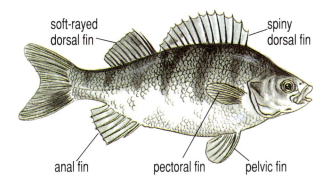

soft-rayed dorsal fin · spiny dorsal fin · anal fin · pectoral fin · pelvic fin

fish ► page 48

fish farming ► **mariculture**

fish trap *noun*
A fish trap is a net, basket or fence, used to catch fish near the shore. Funnel-shaped fences or nets trap fish as they move with the **ebb tide**. Fishermen may splash and make a noise to frighten fish into circular nets suspended from small boats. Bundles of wood which are placed on the sea-bed will also attract some of the fish that dwell there.
At low tide the fishermen waded out to inspect their fish traps.

fishery *noun*
A fishery is an area which supplies large numbers of fish for commercial use. Inland fisheries include lakes, rivers and **fish farms**. Most of the world's **commercial fish** come from sea fisheries.
Species of herring, flatfish, cod, tuna and sardines make up the world's most important fishery sources.

fishing ground *noun*
A fishing ground is a place in the sea where fish are caught.
The rich fishing grounds of the Grand Banks attract many foreign fishing vessels.

fishing industry ► page 50

fishmeal *noun*
Fishmeal is the name given to fish that has been processed. It is added to animal feed as a source of protein or used as a **fertilizer** for crops.
Forty per cent of the world's fish catch is processed into fishmeal.

fjord *noun*
A fjord is a deep, narrow inlet of the sea. It has steep mountain slopes on each side. Fjords were formed when a valley was deepened by a **glacier** and was then flooded by the sea.
They went on a cruise along the Norwegian coast to see the spectacular fjords.

flag ► page 51

flag of convenience *noun*
A flag of convenience is a flag carried by a ship. It is the flag of a country which is not the country in which the ship's owners live. Sometimes, ships register in other countries to avoid strict local regulations.
The ship that ran aground was flying a flag of convenience.

flagellate *noun*
A flagellate is a **microscopic** creature, which has **flagella** to propel itself along.
Many flagellates spin round as they swim.

flagellum (plural **flagella**) *noun*
A flagellum is a thin, hair-like structure found in some **microscopic** organisms. The flagellum waves around to push the creature through the water.
Flagellates use their flagella to swim among plankton.

fish   *noun*

A fish is a **vertebrate** which lives its entire life in the water. Fish make up four classes of animals and scientists have named nearly 22,000 **species**. Fish breathe under water through their **gills**. A fish's scaly body is tapered at both ends and it has fins that help it swim. Many fish have skeletons made of bone. These are called **bony fish**. They contain a **swim bladder**.

Fish with skeletons made of cartilage are called cartilaginous fish.

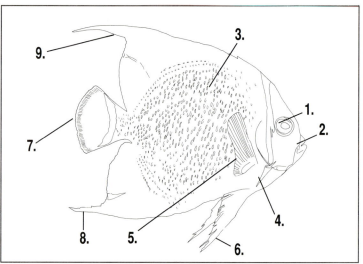

1. eye
2. mouth
3. scales
4. gill cover
5. pectoral fin
6. pelvic fins
7. caudal fin
8. anal fin
9. dorsal fin

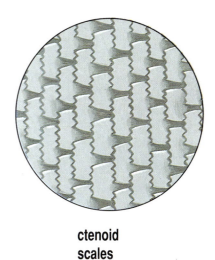

ctenoid scales

cycloid scales

Scales provide body armour for nearly all fish. Most fish have either ctenoid or cycloid scales, both of which are thin and flexible.

Fish breathe by taking mouthfuls of water and then squirting it out through their gills. Gills are feathery organs which are enclosed in pouches on each side of the fish's head. Oxygen from the water passes into the gills. Fish need oxygen to provide energy.

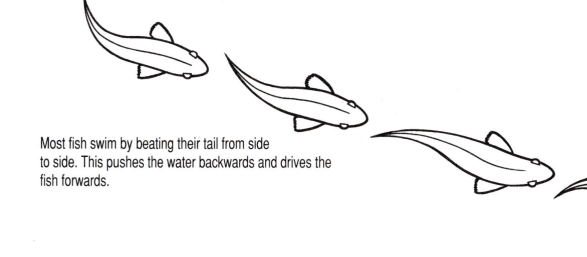

Most fish swim by beating their tail from side to side. This pushes the water backwards and drives the fish forwards.

fishing industry *noun*

The fishing industry is the catching, processing and marketing of fish, shellfish, **seaweeds** and other products from the **sea**, or from inland **fisheries**. There are several different ways of catching fish. These involve the use of large nets, such as **drift nets**, **purse seine nets** and other trawl nets.

The fishing industry employs about six million people worldwide.

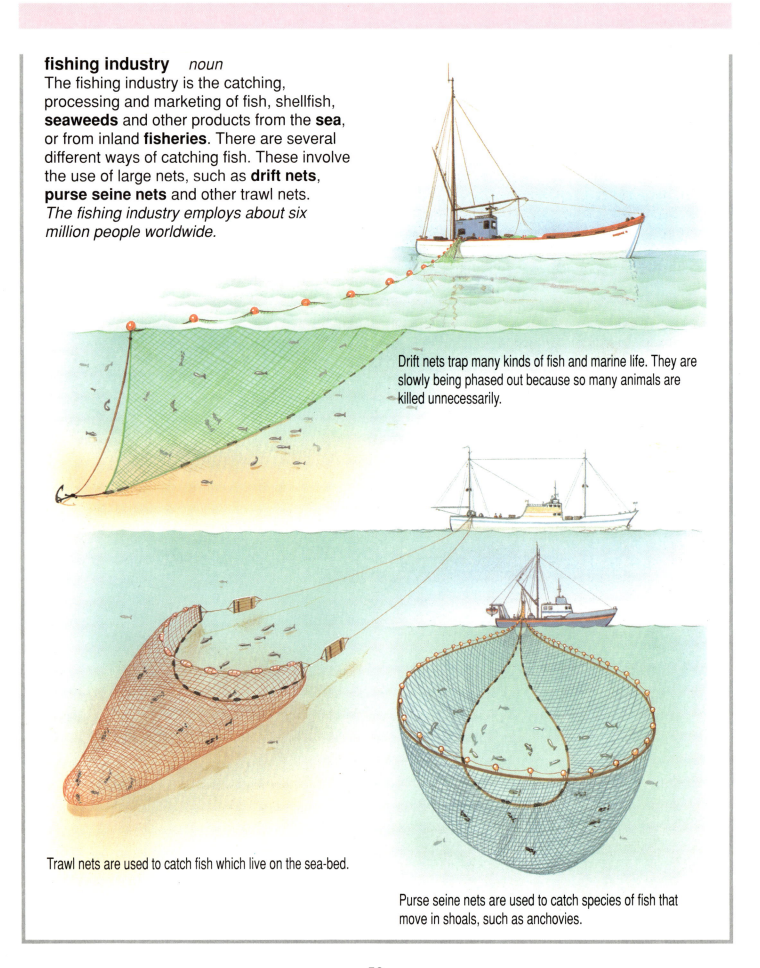

Drift nets trap many kinds of fish and marine life. They are slowly being phased out because so many animals are killed unnecessarily.

Trawl nets are used to catch fish which live on the sea-bed.

Purse seine nets are used to catch species of fish that move in shoals, such as anchovies.

flag *noun*

A flag is a piece of cloth flown from the mast of a ship. It is used to give information to other ships. Most ships carry a flag in the colours of the country in which they are registered. Flags are also used to signal. There is a different flag for each letter of the alphabet, and for each numeral. There are also flags that warn of high winds and other dangers.

Before radios and telephones were invented, flags were the only way to signal between ships.

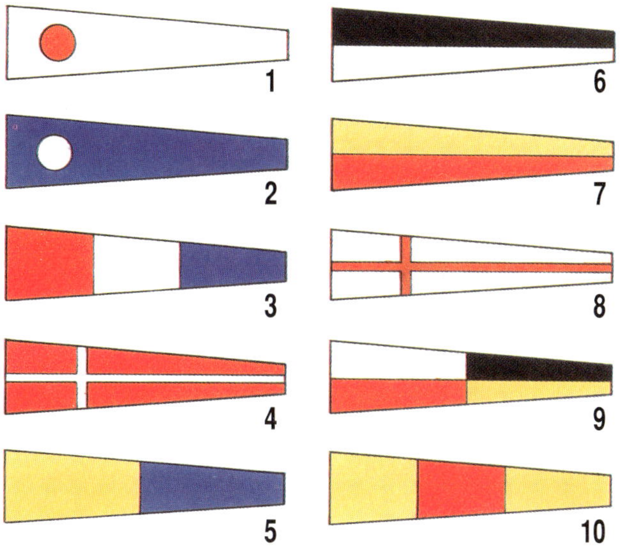

1
2
3
4
5
6
7
8
9
10

A B C D E F G H I

J K L M N O P Q R

S T U V W X Y Z

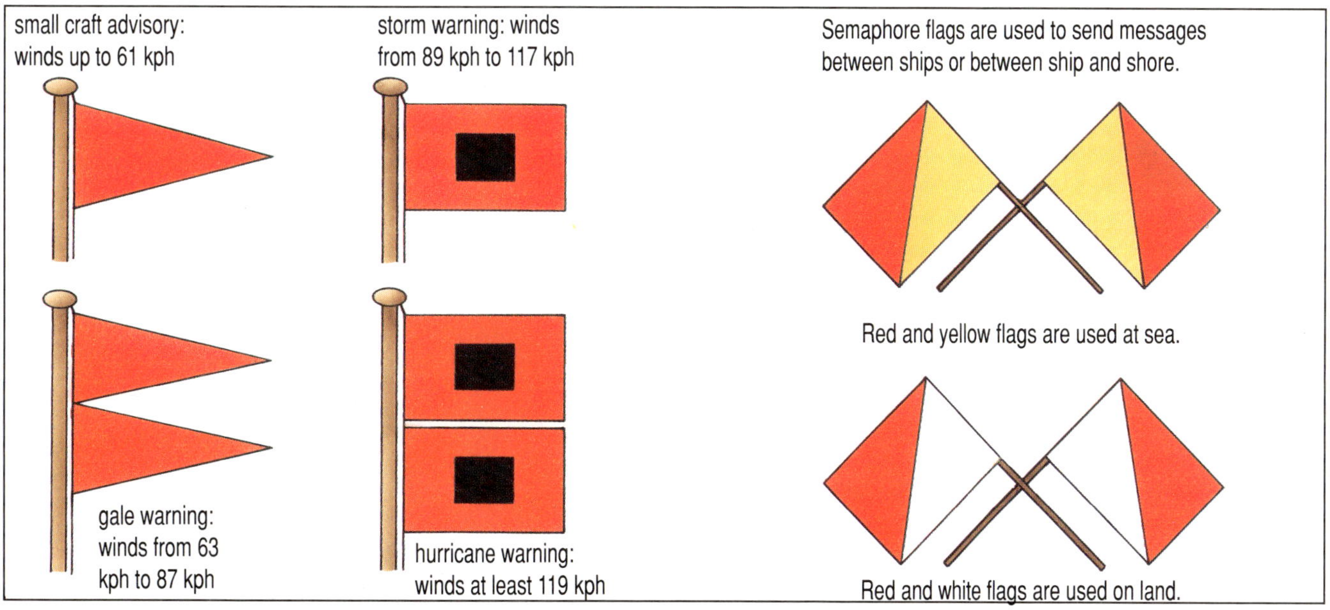

small craft advisory: winds up to 61 kph

storm warning: winds from 89 kph to 117 kph

gale warning: winds from 63 kph to 87 kph

hurricane warning: winds at least 119 kph

Semaphore flags are used to send messages between ships or between ship and shore.

Red and yellow flags are used at sea.

Red and white flags are used on land.

flatfish ► page 54

flipper   *noun*
A flipper is a paddle-like limb. Marine
vertebrates such as seals, **cetaceans**,
penguins and turtles have flippers.
The turtle dragged itself up the beach on its
flippers.

flood current   *noun*
A flood current is a **current** that develops as
the **tide** is rising.
As the tide came in, the yachts had to sail
against a flood current in the estuary.

flood tide   *noun*
Flood tide describes the incoming tide.
The opposite of flood tide is **ebb tide**.
Planks of wood from the wreck were swept
up the beach on the flood tide.

Florida Current   *noun*
The Florida current is a **surface current** that
flows up the east coast of Florida, in the
United States of America, where it mixes
with the **Gulf Stream**.
The Florida Current carries warm water from
the Caribbean to the beaches of Florida.

flotsam   *noun*
Flotsam describes an object or objects found
floating in the sea, such as wood or cargo. It
may come from a wrecked ship, or it may
have been thrown overboard from a vessel
in distress. Maritime law states that flotsam
remains the property of its original owner, no
matter how long it lies in the sea.
The water in the bay is covered in flotsam
from last night's shipwreck.

flounder ► **flatfish**

flow   *verb*
Flow describes the movement of the **tide**
towards the **shore** or up an **estuary**.
The opposite of flow is **ebb**.
The boat could not sail out of the estuary
until the tide began to flow.

flying fish   *noun*
A flying fish is a silvery fish with large
pectoral **fins** that can be spread out like
wings. Flying fish swim rapidly towards the
surface of the water. Then they flick their
powerful tail to leap out of the water and
glide through the air. There are about 50
different species of flying fish.
Flying fish are common in warmer parts of
the ocean.

fog   *noun*
Fog is a kind of thick mist. It looks like a
cloud over the surface of the ground or sea.
Fog is made up of lots of water droplets in
the air.
Fog forms when air containing a lot of
moisture is cooled, and the water vapour
turns into droplets.

foghorn   *noun*
A foghorn is a horn that sounds in foggy
weather to warn ships to stay clear of
dangerous rocks.
The foghorn boomed a warning out to sea.

food chain   *noun*
A food chain is the name given to a group of
plants and animals which feed upon each
other in an **ecosystem**. At the bottom of the
chain are plants or the **algae** of the
phytoplankton. These produce their own
food by **photosynthesis**. Next come
animals which feed on plants, then animals
which feed on other animals.
Killer whales are at the top of the ocean food
chain.

food pyramid *noun*
A food pyramid describes the weight, or mass, of **organisms** at each level of a **food chain**. When an animal eats a plant or another animal, some of the food is wasted as undigested food and body waste. More food is burned up to provide energy for bodily processes and movement. This means that the total weight of all the organisms at one level of the food chain is less than the weight of those at the level below.
The top layer of a food pyramid contains less energy than the lower layers.

food web *noun*
A food web is a series of **food chains** linked together. There are often many different kinds of living organism at each level of a food chain. When they are linked together, they form a food web.
The ocean food web is very large and complex.

foraminiferan *noun*
A foraminiferan is a one-celled animal. It is no larger than five millimetres across. Some foraminiferans are too small to see without a microscope. They live on the sea-bed or among **plankton**. Foraminiferans have shells of calcium carbonate which often form several round chambers. They have tiny, soft, branching arms which reach out and trap even tinier water creatures to eat.
Large areas of the ocean floor are covered in the chalky remains of foraminiferans.

Forties *noun*
The Forties describes the part of the ocean between the **latitudes** of 40 degrees and 50 degrees north and south, where the **prevailing winds** blow from the west. They are often called the Roaring Forties because they have many strong winds and gales all year round.
Many of the sailors were seasick as their ship sailed through the rough seas of the Forties.

fossil *noun*
A fossil is the remains of a living thing, or organism, that lived on the Earth millions of years ago. Fish fossilized well because their skeletons were made of hard bone. Fossils are usually found in sedimentary rocks.
Fossils of bivalves are common in limestone rocks that were formed on the ocean floor.

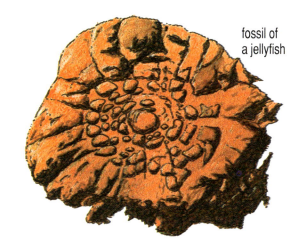

fossil of a jellyfish

fouling *noun*
Fouling describes the growth of **algae**, **barnacles** and other marine **organisms** on the bottom of boats. A special paint is sometimes painted on the bottom of boats called anti-fouling, which releases poisons in order to kill the fouling.
They painted the boat with a special paint to stop fouling.

French angelfish ► **butterfly fish**

French grunt ► **grunt**

freighter ► **cargo ship**

fresh water *noun*
Fresh water is water which does not contain much salt. It is easily turned into drinking water. The water of most rivers and lakes is fresh water. Less than one per cent of the Earth's water is fresh water. Fresh water can be made by taking salt out of sea water. This process is called **desalination**.
In an estuary, fresh water mixes with sea water and becomes salty.

flatfish *noun*

A flatfish is a wide, flat fish, with two long fins, like fringes, at its sides. In fact, the fish is lying on its side. When a flatfish swims, it looks as if it is bending from side to side. young flatfish look like ordinary fish, but as they grow bigger, they become flatter. One eye moves to the other side of the head, the mouth twists, and the fish starts to lie on one side. Plaice, sole, halibut, turbot and flounder are all flatfish.

The flatfish was lying on the sea-bed.

plaice

brill

turbot

halibut

flounder

lemon sole

frigate *noun*

A frigate is a kind of warship. It has guns and missiles on deck and sometimes carries a helicopter.
Frigates have radar and sonar to detect enemy aircraft, ships and submarines.

frigatebird *noun*

A frigatebird is a large **seabird** with long wings and a forked tail. Frigatebirds behave like pirates, and are often called man-o'-war birds. They rob other seabirds of their fish in mid-air and also steal their eggs and chicks. Frigatebirds also catch their own fish. In the breeding season, the male bird develops a huge, crimson throat pouch which he inflates like a balloon, to attract a mate.
A frigatebird bird swooped down and snatched a flying fish from the air.

front *noun*

A front occurs when two air masses of different temperatures meet.The front edge of a moving mass of warm air is called a warm front. The front edge of a mass of cold air is a cold front. A front is also where two water masses of different temperatures or different salt content meet in the ocean.
The weather became milder as the warm front approached.

fry *noun*

Fry is the name given to a very young fish.
The stickleback fry stayed in the shelter of the weed out of sight of predators.

fulmar *noun*

A fulmar is a seabird belonging to the **petrel** family. Fulmars breed on rocky shores. They particularly like to feed on fatty substances such as whale blubber.
Fulmars lay only one egg in their nest.

56

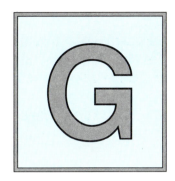

gannet   *noun*
A gannet is a large seabird, with long wings, a large bill and a long, wedge-shaped tail. Gannets are **diving birds** and catch fish by plunge-diving into the sea.
Young gannets and boobies wander the ocean for several years before starting to breed.

gas field   *noun*
A gas field is a place where natural gas is found.
They drilled for gas in the huge gas field in Australia.

gastropod ► page 58

giant clam   *noun*
A giant clam is a huge **bivalve** which can reach up to 135 centimetres in length. Giant clams are thought to live for several hundred years. By day, the clam opens to expose a large, fleshy mantle to the light. It does this because **algae** living inside the mantle need light for **photosynthesis**. This is an example of **symbiosis**.
Some of the largest giant clams may be hundreds of years old.

giant squid   *noun*
The giant squid is the largest **invertebrate** in the world. It grows up to 20 metres long, including the **tentacles**. It can weigh over two tonnes. Giant squids live in deep water in the **Atlantic Ocean**, and in the Pacific Ocean near New Zealand.
The sperm whale is probably the only predator of the giant squid.

giant squid

gill   *noun*
Gills are tiny, branching, finger-like structures used for breathing by many underwater animals. The feathery shape gives a large area for absorbing oxygen from the water. In fish and most molluscs and **crustaceans**, the gills are inside the body and water is pumped over them. Some aquatic worms and larvae have gills that stick out into the water.
The mussel opened its shells a little as it wafted water over its gills.

gill cover   *noun*
A gill cover is a flap of bone which protects the gills of **bony fish**.
Sharks, rays and other cartilaginous fish do not have gill covers.

gill net   *noun*
A gill net is a net which traps a fish by its gill covers. As the fish tries to back out of the net, the gill covers catch in the mesh. The size of the mesh depends on the kind of fish to be caught. The nets can be placed at any depth in the sea by means of floats and anchor cables.
The sea turtle was trapped in a gill net and drowned.

gastropod *noun*

A gastropod is a **mollusc**. Gastropods have **tentacles** on their head for tasting and touching. They may have a single shell, or no shell at all. Their eyes are on tentacles. Gastropods creep along on a large, flat foot. Most of them feed using a strip of small, horny teeth, called a radula. Some gastropods are predators, while others feed on **plankton** or **detritus**.

Snails, sea slugs, sea hares and limpets are all gastropods.

top shell

turret shell

cowrie shell

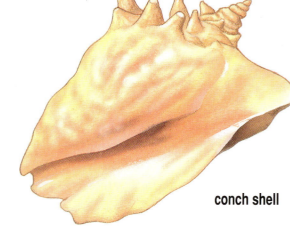

conch shell

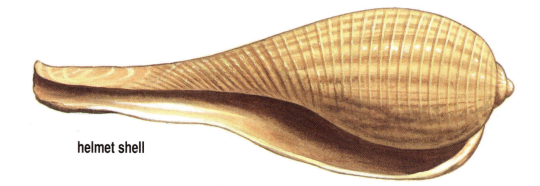

helmet shell

glacier   *noun*
A glacier is a river of ice. It is formed from very deep snow packed down so hard that it turns to ice. The weight of the ice makes the glacier move downhill. Most glacier range in thickness from about 100 to 3,000 metres.
Most glaciers move only a few centimetres a day.

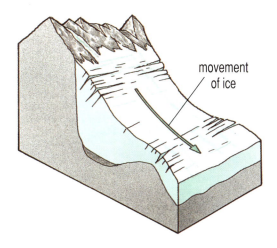

movement of ice

global ocean   *noun*
The global ocean describes all the world's oceans put together. The oceans are all connected to each other by water.
All humans are affected by the global ocean and its effects on the weather.

global warming   *noun*
Global warming is the increase in the average temperature of the Earth's atmosphere due to the **greenhouse effect**. Scientists expect the increasing temperatures to upset the world's climate and raise the sea-level.
If the polar ice melts as a result of global warming, the sea-level will rise.

goby (plural **gobies**)   *noun*
A goby is a small fish that lives in shallow water along **coasts**, **coral reefs** and **mangrove swamps**. Gobies have two dorsal fins, and the pelvic fins are joined together to make a single fin. In some gobies, this fin forms a sucker for clinging to rocks.
Gobies are about 102 millimetres long.

Grand Banks   *noun*
The Grand Banks is part of the North American **continental shelf** that lies to the north-east of Newfoundland. It is an important **fishing ground**, especially for cod.
Boats from all over the world come to fish on the Grand Banks.

gravel   *noun*
Gravel is small pieces of rock between two and four millimetres across. Gravel collects on **beaches** where strong **tidal currents** wash **sand** out to sea.
Gravel comes between sand and pebbles in size.

gravity   *noun*
Gravity is a force which pulls objects towards the centre of the Earth. The larger and denser a star or a planet is and the nearer it is to other objects the more strongly it pulls them towards it.
The pull of the Earth's gravity makes dead organisms sink to the ocean floor.

Great Barrier Reef   *noun*
The Great Barrier Reef is the world's largest **coral reef**. It lies off the east coast of Australia. The Great Barrier Reef is 2,012 kilometres long, and is large enough to be seen from the Moon. There are about 400 species of **polyps**, about 1,500 species of **fish** and many kinds of birds and **crabs**, **sea turtles** and **giant clams** living on the Great Barrier Reef.
Many tourists visit Queensland, Australia, to see the Great Barrier Reef.

green seaweed   *noun*
Green seaweed is a small, delicate seaweed. There are over 5,000 different kinds of green seaweed. They may form thin sheets, flat ribbons or branching, tree-like shapes. Sea mosses are flimsy and feathery. Mermaid's cup has tiny cups on slender stalks like small, green toadstools.
The sea lettuce sold as food and medicine in China is really a green seaweed.

greenhouse effect *noun*
The greenhouse effect is a theory that explains why the Earth's atmosphere is slowly warming up. Scientists think that the greenhouse effect is caused by the build-up of the gases **carbon dioxide** and methane in the atmosphere. The gases prevent heat escaping from the lower part of the atmosphere.
Some scientists think that the greenhouse effect will raise average temperatures by up to five degrees Celsius by the year 2050.

Greenpeace *noun*
Greenpeace is an organization that campaigns to protect the world's environment and wildlife from **pollution** and destruction.
Small boats belonging to Greenpeace sailed between the whalers and the whales.

grouper *noun*
A grouper is a kind of **fish**. It is a large sea bass. Some of the biggest groupers are up to four metres long and can weigh over 300 killogrammes. They lie in wait for other fish between rocks or in coral caves. Groupers move slowly, and are easily caught with fishing rods or spears.
The spear fishermen were looking for groupers along the edge of the reef.

groyne *noun*
A groyne is a small, wooden **breakwater**. Groynes are built at intervals along a **beach** to stop **sand** and **shingle** being washed away.
Banks of shingle had piled up against the groynes.

grunion *noun*
A grunion is a silvery fish that **spawns** on the beaches of the Pacific coast of North America. Huge **shoals** of grunion invade the beaches after very **high tides** and lay their eggs there. The eggs stay hidden in the sand until the next very high tide causes them to hatch.
The beach was alive with millions of wriggling grunion.

grunt *noun*
A grunt is a colourful tropical fish with a flattened body. It makes a grunting sound by grinding together certain teeth at the back of its mouth.
Several grunts were waiting for the services of a cleanerfish.

guano *noun*
Guano is the name given to large deposits of the droppings of birds, seals or bats. It is collected by people and makes a valuable **fertilizer**.
There are islands off the coast of Peru which are white with guano from the seabirds that nest there.

guillemot ► **auk**

Guinea Current *noun*
The Guinea Current is a warm **ocean current** that flows down the west coast of Africa, from Guinea to Cameroon.
Between the cold Canary and Benguela Currents are the warm waters of the Guinea Current.

gulf *noun*
A gulf is an area of sea which is almost enclosed by land.
There are many ports along the sheltered coast of the gulf.

Gulf of Alaska *noun*
The Gulf of Alaska is part of the south coast of Alaska.
Many sea otters live in the Gulf of Alaska.

Gulf of Arabia ► **Persian Gulf**

Gulf of Bothnia *noun*
The Gulf of Bothnia is a **gulf** of the Baltic Sea between Sweden and Finland.
Parts of the Gulf of Bothnia freeze in winter.

Gulf of California *noun*
The Gulf of California, or Sea of Cortes, is an arm of the Pacific Ocean. It lies between the Californian peninsula and mainland Mexico.
The Gulf of California is about 1,100 kilometres long.

Gulf of Mexico *noun*
The Gulf of Mexico is an area of sea off the south-east coast of the United States of America. It is surrounded by the coasts of the southern United States, north-east Mexico and Cuba. The Gulf is connected to the **Atlantic Ocean** by the Straits of Florida, and to the **Caribbean Sea** by a narrow channel between the Yucatan Peninsula and Cuba.
There are many oilfields off the shores of the Gulf of Mexico.

Gulf Stream *noun*
The Gulf Stream is a large, warm, **ocean current**. It flows from the **Gulf of Mexico** up the east coast of the United States, then north-east to the middle of the North **Atlantic Ocean**. Here, it joins the **North Atlantic Drift** and the Canary Current.
The mild climate of north-west Europe is partly due to the warming effect of the Gulf Stream.

gull ► page 62

guyot *noun*
A guyot is a flat-topped, underwater mountain. It forms over thousands of years as waves wear away, or erode, the top of a **volcanic island**. This then sinks below the water as the **ocean floor** subsides.
Fossil corals have been found on the tops of guyots 2,000 metres below the ocean's surface.

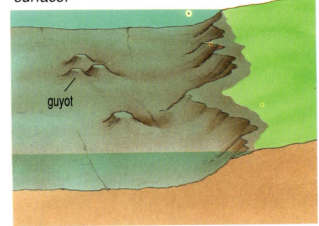

guyot

gyre *noun*
A gyre is a circular **ocean current**, several hundred kilometres across. Gyres are often caused by the **Coriolis force**.
Water in the ocean gyres flows clockwise in the northern hemisphere, and anti-clockwise in the southern hemisphere.

gull *noun*

A gull, or seagull, is a sturdy, noisy **seabird**. It has long, pointed wings, webbed feet and a strong, slightly hooked beak. Most gulls are grey and white in colour, but some have dark heads. Gulls often scavenge along the **strandline**, and on fields and rubbish tips. They catch fish and **invertebrates**, and steal the eggs and chicks of other birds. Gulls nest in large colonies.

The gulls were following the ship, waiting to catch any scraps of waste thrown overboard.

black-backed gull with chick

herring gull

herring gull egg

ivory gull

kittiwake

Californian gull

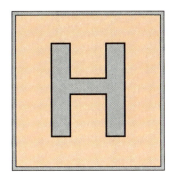

habitat *noun*
A habitat is a place where a particular plant or animal, or a community of plants and animals, lives. A habitat includes physical features such as rocks, soil and water, animals and plants, and climate.
A habitat may be very large, like the ocean, or very small, like a pond.

haddock ▶ **cod family**

hagfish *noun*
A hagfish is a **jawless fish** that lives on the sea-bed. It feeds mainly on dead and dying fish. Hagfish have no visible eyes. Instead, they use **barbels** around their mouth to find their **prey**. They have a rasping tongue covered in horny teeth.
Hagfish have several sets of hearts.

hake *noun*
A hake is a large, narrow fish which looks similar to a cod. Hakes have big, sharp teeth for hunting fish and **squid**. They have two dorsal fins, and a ribbon-like anal fin.
The eggs of hake contain oil droplets that help them float on the water until they hatch.

halibut ▶ **flatfish**

halophyte *noun*
A halophyte is a plant that can live in salty soil. Many **saltmarsh** plants are halophytes. Halophytes are often succulent. This means they have thick leaves and stems in which to store water.
Only halophytes can grow on the muddy shores of an estuary.

hammerhead *noun*
A hammerhead is a kind of **shark**. It has a large, hammer-shaped head with eyes at the outer edges of the 'hammer'. Scientists think this large head may contain organs for picking up faint pulses of electricity coming from the shark's **prey**.
A group of hammerhead sharks moved above the divers.

harbour *noun*
A harbour is a place where ships can moor, or anchor. It may be a natural cove or an artificial harbour created in the shelter of a **jetty** or **breakwater**.
The sailors were relieved to reach the harbour before the storm arrived.

harpoon *noun*
A harpoon is a spear which has backward-curving hooks. It is used to kill large fish, **whales** and other big marine animals. Fishing harpoons are usually thrown by hand. Whaling harpoons are fired from a gun. The hooks prevent the harpoon coming out of the animal's flesh.
The whaler had a harpoon mounted in the bow of the ship, ready to fire at the passing whale.

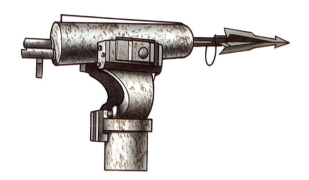

hatchery *noun*
A hatchery is a pond or cage where the eggs of fish, shellfish or other animals are kept under special conditions to encourage them to hatch. This is often done when the eggs are in danger from predators.
The rare turtle eggs were taken to a hatchery.

hatchetfish *noun*

A hatchetfish is a small, silvery **deep sea fish** with a very deep, thin body. Hatchetfish have tubular eyes that face upwards so they can see the silhouettes of **crustacean** prey outlined against the light. They have **light organs** on their abdomen to prevent their own **predators** seeing their shadow. Hatchetfish is also the name of a small freshwater fish that can fly out of the water.
The mirror-like sides of hatchetfish reflect the dark sea around them.

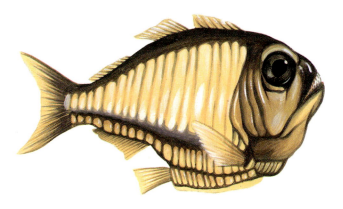

headland *noun*

A headland is a narrow piece of land which juts out into the sea. A headland forms where there are bands of soft and hard rock along the **coast**. The sea wears away the soft rock faster than the hard rock, forming a series of headlands and **bays**.
There was a lighthouse at the end of the headland that warned ships of the hidden rocks.

heavy metal *noun*

A heavy metal is a metal that has a large weight. The most common are iron, lead, copper, zinc, nickel and cadmium. Many of these metals are poisonous when they find their way into water supplies. They may be released into rivers and seas in the form of industrial waste. Here, heavy metals can pollute drinking water and kill underwater plants and animals.
The dead fish washed up on the shores of the estuary had been poisoned by heavy metals from the nearby steel works.

herbivore *noun*

A herbivore is an animal that feeds on plants or on **algae**.
The algae growing on the reef are grazed by herbivores such as sea slugs and damselfish.

hermit crab *noun*

A hermit crab is a crab that lives in the cast-off shells of large **gastropods**, or **sea snails**. Unlike other crabs, the hermit crab has a soft abdomen and needs the protection of a shell. When danger threatens, the hermit crab retreats into its shell and closes the opening with a large claw. The borrowed shell is also tough enough to withstand buffeting by the **waves** on the **beach**.
The hermit crab had grown too big for its shell and was looking for a larger one.

herring *noun*

The herring is a silvery fish with a narrow body that tapers to a thin ridge along the abdomen. Herrings have a single dorsal fin, and their pelvic fins are set well back on the abdomen. Their scales are easily rubbed off. Herrings form huge **shoals** in the open sea and along the coast. The herring family includes important commercial fish such as herrings, sardines, shads, sprats, pilchards and menhaden.
The large shoals formed by fish of the herring family are easily found by echo-sounding.

high pressure area   *noun*
A high pressure area is an area where the **atmospheric pressure** is high. It is the opposite of a **depression** or **cyclone** and is often called an anticyclone. Winds blow around a high pressure area in a clockwise direction in the northern hemisphere, and in an anti-clockwise direction in the southern hemisphere.
The weather was dry and clear as a high pressure area settled over the country.

high water mark   *noun*
The high water mark is the highest part of a beach which can be reached by a **spring tide**. It is often marked by a line of seaweed called the **strandline**, which is left behind by the tide.
Animals that live below the high water mark must avoid being washed away by the tide.

holdfast ► **seaweed**

horse latitudes   *plural noun*
Horse latitudes are belts of **high atmospheric pressure** that circle the Earth between latitudes 30 to 35 degrees north and south of the **Equator**. They are areas of dry downward moving air which produce arid climates on land. At sea, horse latitudes are called the calms of Cancer in the north and the calms of Capricorn in the south.
They lost the race because the sailing ship was trapped in the calm air of the horse latitudes.

horseshoe crab   *noun*
A horseshoe crab, or king crab, is an **invertebrate**. It is not a true crab, but is related to the spider. The body of a horseshoe crab is in three parts. There is a hard, shield-like front section, a smaller middle section, and a long, sharp tail spine. The mouth is on the lower surface of the body, and is surrounded by five pairs of legs.
Horseshoe crabs push through the mud on the sea-bed, feeding on worms and molluscs.

hot spot   *noun*
A hot spot is part of the Earth's **mantle** where molten rocks called **magma** rise upwards through the **crust**. A **volcano** forms where a hot spot reaches the surface of the Earth.
The volcanic island of Hawaii lies over a hot spot in the Earth's mantle.

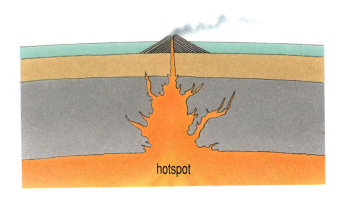

hotspot

hot spring ► **hydrothermal vent**

hovercraft   *noun*
A hovercraft is a vehicle that floats on a cushion of air. The cushion of air is made by pumping air downwards. The air is then kept in by a large curtain of rubber. Aircraft propellers on top of the hovercraft propel it forwards. Hovercraft are used as ferries. They can travel on land, water and ice.
They took the car across the English Channel to France on the hovercraft.

Hudson Bay   *noun*
The Hudson Bay is a huge bay in the north of Canada. It is bordered by the states of Manitoba, Ontario and Quebec. The Hudson Bay is linked to the **Atlantic Ocean** by the Hudson Strait.
The Foxe Channel links the Hudson Bay to the waters which lead to the Arctic Ocean.

Humboldt Current   *noun*
The Humboldt Current is a large, cold **ocean current** that flows northwards up the west coast of South America.
The waters of the Humboldt Current support a large fishing industry.

hurricane *noun*

A hurricane is the strongest possible **wind**. It measures force 12 on the **Beaufort Scale**. A hurricane is caused by a very strong **depression**, or a revolving system of winds and thunderstorms. In the centre of the hurricane is an area of calm, called the eye. After the strong wind comes torrential rain, followed by the calm of the eye, then the wind comes back from a different direction.
The wind dropped as the eye of the hurricane passed over.

hydrographer *noun*

A hydrographer is a person who describes, measures and maps the Earth's waters, including the **oceans**, seas, lakes, rivers and streams. Hydrographers chart **currents** and **tides** and the shape of the **ocean floor**.
The maps made by hydrographers help sailors to navigate.

hydrological cycle ► water cycle

hydrostatic pressure *noun*

Hydrostatic pressure is the pressure of water at a given **depth**. It depends upon the weight of the water **mass** above that depth. Hydrostatic pressure is usually measured in atmospheres. One atmosphere is a pressure of 1.033 kilogrammes per square centimetre, which is the average weight of the atmosphere at **sea-level**.
At a depth of 10,000 metres, the hydrostatic pressure is about 1,000 atmospheres.

hydrothermal vent *noun*

A hydrothermal vent is a hot spring on the deep **ocean floor**. It is formed where water leaks through cracks in the ocean floor. The water is warmed by the hot rocks below and then seeps back to the sea-bed. Minerals such as iron, nickel, silver, copper and sulphides dissolve in the hot water and are laid down, or deposited, around the spring.
Tube worms that are three metres long live in hydrothermal vents.

hydrozoan *noun*

A hydrozoan is a **cnidarian**. It has two different shapes during its life cycle. One shape is an anemone-like **polyp**, and the other is a **medusa** that looks like a small **jellyfish**. Many hydrozoans are branching colonies of **polyps** attached to rocks or seaweeds. When they breed, they bud off a medusa. These swim away and produce new polyps. Some hydrozoans, such as the **Portuguese Man-of-War**, form large, floating colonies.
The medusae of many hydrozoans live in plankton.

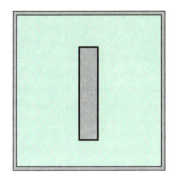

ice cap *noun*

An ice cap is a thick covering of ice on an area of land. As the snow builds up in layers, the increasing weight squeezes, or compresses it into ice. Large ice caps are sometimes called ice sheets.
The South Pole lies on the Antarctic ice cap.

ice floe *noun*

An ice floe is a flat sheet of floating ice. Ice floes break away from ice shelves when the sea warms up in spring.
Ice floes vary in size from a few metres to several hundred metres across.

ice sheet ► ice cap

ice shelf *noun*

An ice shelf is a large, flat mass of thick floating ice that is attached to the coast. The shelf may also be attached to the sea-bed and to offshore islands. It is fed with ice by **glaciers** or **ice caps**, and grows slowly out into the sea. Large, flat icebergs break off from the outer edge of the shelf.
The Ross Ice Shelf in Antarctica is about the size of France.

iceberg *noun*

An iceberg is a large piece of ice which floats in the sea. Icebergs break off from the edges of **ice sheets** and **glaciers** where they meet the sea. Most Antarctic icebergs have broken off from **ice shelves**, and are flat-topped. Arctic icebergs usually come from glaciers and have jagged shapes.
Some icebergs are as tall as a ten-storey building.

icefish *noun*

Icefish, or 'white-blooded fish', are a group of **fish** that have colourless blood, which gives them a rather ghostly appearance. Some icefish are active **predators**, while others lie in wait for their **prey** on the sea-bed.
Several kinds of icefish live in Antarctic waters.

ichthyosaur *noun*

An ichthyosaur was a prehistoric marine reptile that lived between 255 and 65 million years ago. The ichthyosaur looked rather like a **dolphin**, but it had a longer, narrower snout and its tail **fins** were vertical, not horizontal.
Ichthyosaurs swam by flexing their body and tail, just as fish do.

Indian Ocean *noun*

The Indian Ocean is a large area of sea. It lies between Africa, Asia, Australia and Antarctica.
The Maldive Islands lie in the Indian Ocean.

inshore rescue boat *noun*

An inshore rescue boat is a small, inflatable rubber boat which is used to rescue people in difficulty near the shore. It is very fast and can go into much shallower water than a **lifeboat**. It is also quick and easy to launch.
The helicopter directed the inshore rescue boat to the struggling yachtsmen.

Intergovernmental Oceanographic Commission (IOC)   *noun*

The Intergovernmental Oceanographic Commission, or IOC, is an international organization. It promotes the international scientific investigation of the oceans.
The IOC helped to produce a world map of the ocean floor.

International Hydrographic Bureau (IHB)   *noun*

The International Hydrographic Bureau, or IHB, is the headquarters of the International Hydrographic Organization (IHO). This organization helps different countries to exchange information about oceanography. It also helps them to agree on **navigation** methods and charts.
Echo-sounding data and satellite information are collected by the IHB in order to produce bathometric charts.

International Whaling Commission (IWC)   *noun*

The International Whaling Commission, or IWC, is an organization that tries to prevent **overfishing** of whales by encouraging research and regulation. In 1986, the IWC banned all commercial whaling to let numbers recover. Membership of the IWC is voluntary.
Some member countries are now threatening to break away from the IWC in order to start whaling again.

intertidal zone   *noun*

The intertidal zone, or littoral zone, is part of the **coast**. It lies between the **high water mark** and the **low water mark**.
In most parts of the world, the intertidal zone is flooded by the tide twice a day.

invertebrate   *noun*

An invertebrate is an animal without a backbone.
Many small invertebrates, such as anemones, crabs and sea urchins, live in rock pools.

Ionian Sea   *noun*

The Ionian Sea is the deepest part of the **Mediterranean Sea**. It lies between the south-east coast of Italy and the west coast of Greece.
The Ionian Islands lie in the eastern part of the Ionian Sea.

Irish Sea   *noun*

The Irish Sea lies between England and Ireland. It is connected to the **Atlantic Ocean** through channels to the north and south. The Irish Sea is about 370 kilometres long and is 225 kilometres wide at its widest point. The Isle of Man lies in the Irish Sea.
The English visitors were seasick after their rough crossing of the Irish Sea.

Irminger Current   *noun*

The Irminger Current is a branch of the warm **North Atlantic Current**. It curves westwards, past the south coast of Iceland, and then flows south-west into the Labrador Sea.
Near Iceland, the warm water of the Irminger Current mixes with cold, nutrient-rich Arctic water.

island   *noun*

An island is an area of land that is surrounded by water. An island may be only a few metres across, or it may be many thousands of kilometres wide. The largest island on Earth is the continent of Australia. A group of islands is called an **archipelago**.
They spent their holiday on a cruise round the Greek islands.

island arc   *noun*
An island arc is a chain of **volcanic islands**. Island arcs form when one part of the **ocean floor** is pushed underneath another part. This happens where two **tectonic plates** meet at an ocean trench. At a certain depth, the rocks of the ocean floor melt, and the molten rock rises through the crust to form a string of volcanoes.
The Aleutian Islands are part of an island arc not far from the Aleutian Trench in the North Pacific Ocean.

isopods   *noun*
Isopods are a group of **crustaceans** with long, slightly arched bodies covered in armour-like plates. They include woodlice, pill bugs and sea slaters. Many isopods are **parasites** of other marine animals.
The lobster pot was empty because a group of isopods had eaten the bait.

isthmus   *noun*
An isthmus is a narrow strip of land which joins together two large land masses. There is sea on either side of an isthmus.
The Isthmus of Suez links Africa and Asia, and lies between the Mediterranean and Red Seas.

Isthmus of Panama   *noun*
The Isthmus of Panama is a narrow strip of land that joins North America to South America. It lies between the **Pacific Ocean** and the **Caribbean Sea**.
The Panama Canal is cut across the narrow Isthmus of Panama.

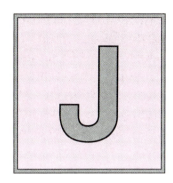

jacks   *noun*
Jacks are a large group of ocean fish. They are rather like mackerel, but have a deeper body and a deeply forked tail on a narrow stalk. Young jacks travel in **schools**, but adult jacks travel alone. **Pompano**, bluefish, pilotfish and yellowtail belong to the jack family.
The fisherman caught several jacks which he took to the market to sell.

Java Sea   *noun*
The Java Sea is part of the **Pacific Ocean**. It lies between the islands of Java and Borneo.
During the ice age, parts of the Java Sea were above sea-level and Asian animals were able to migrate to Indonesia.

Java Trench   *noun*
The Java Trench is a deep ocean trench, found off the south coast of the island of Java.
The deepest part of the Indian Ocean is the Java Trench, which reaches a depth of 7,725 metres.

jawless fish   *noun*
A jawless fish is a **cartilaginous fish** that has no jaws and no proper skull. It has no paired fins, and looks rather like an eel. Its gills open to the outside through a row of holes. **Lampreys** and **hagfish** are both examples of jawless fish. Lampreys live in both salt water and fresh water. Hagfish live only in the sea. There are about 30 species of lamprey and 15 kinds of hagfish.
Jawless fish have slimy, scaleless skins.

jellyfish *noun*
A jellyfish is a **cnidarian**, with a bell-shaped, jelly-like body and long, trailing **tentacles** which have stinging cells. Jellyfish swim through the water by **jet propulsion**, squeezing water out of their body.
The stinging tentacles of jellyfish act as floating traps for other marine organisms.

jet propulsion *noun*
Jet propulsion describes how an object is propelled forwards when a high speed jet of air or water rushes out of the back of the object. Rocket and jet engines work in this way. Animals, such as **jellyfish, octopus** and **squid**, squirt out jets of water to shoot themselves forwards.
The octopus used jet propulsion to make a speedy escape.

jet stream *noun*
Jet streams are twisting bands of high-speed air currents that flow eastwards 10 to 50 kilometres above the Earth. They reach speeds of 500 kilometres an hour, and they have a strong effect on the weather. The position of the jet streams changes with the seasons.
Aided by the jet stream, the plane made a very fast crossing of the Atlantic.

jetsam *noun*
Jetsam is material that is thrown overboard from a ship to make it lighter.
The jetsam was washed up on the beach.

jetty *noun*
A jetty is a structure often built of stone, which is built out into the sea at the entrance to a **harbour** or **estuary**. It protects the mainland against storm waves and movements of **sand** and **mud**. The Columbia River on the Pacific coast of the United States of America has one of the world's longest jetties. It is nearly 7·2 kilometres long.
The family stood on the jetty and waved as he sailed out of the harbour.

John Dory *noun*
A John Dory is a tall, thin **fish** with a large black spot ringed with yellow on each side. In older fish, the first eight to ten rays of the dorsal **fin** develop long trailing tips.
The John Dory moved slowly towards its prey, then suddenly opened its large mouth and sucked in the little fish.

junk *noun*
A junk is a flat-bottomed, Chinese sailing vessel with a high **stern** and an overhanging **prow**. It has large, square sails that can be raised or lowered like Venetian blinds.
In Hong Kong harbour, junks compete for space with luxury yachts.

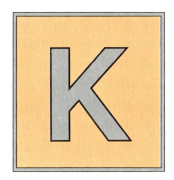

kelp   *noun*
A kelp is a large **brown seaweed** that usually grows below **low water mark**. The fronds of the giant kelp grow over 65 metres long. Air bladders keep the fronds afloat. In California, in the United States of America, people use large machines to harvest kelp and turn it into fertilizer.
Along some coasts, large kelps form underwater forests.

kelp forest   *noun*
A kelp forest is an underwater forest of **brown seaweeds** called **kelps**. Kelps grow in shallow water near the **coast**. The young of many **commercial fish** live in the shelter of kelp forests. **Sea urchins** graze on the kelps, and **abalones** and other **bivalves** thrive there. **Squid** and **cuttlefish** go there to **spawn**.
The sea otters were hunting for abalones in the kelp forest.

killer whale   *noun*
The killer whale, or orca, is a large **dolphin** with a striking pattern of black and white on its body. It can grow up to nine metres long and can weigh up to 725 kilograms. Killer whales travel in **schools** of 40 or more. They are fierce **predators**, and hunt in packs. Killer whales feed mainly on **seals**, sealions, **penguins**, fish and sometimes other whales. They live in all the oceans of the world.
The sealions became agitated as a pack of killer whales appeared close to the shore.

king crab ► horseshoe crab

knot   *noun*
A knot is a measure of speed at sea. One knot is a speed of one nautical mile per hour, or 1,852 metres an hour.
As the wind got up, the yacht increased its speed to 10 knots.

krill   *noun*
A krill is a small, shrimp-like **crustacean**. Huge shoals of krill live in the Arctic and Antarctic Oceans. They are eaten by fish, seabirds and filter-feeding whales.
The humpback whales were gulping in huge mouthfuls of krill, which they then filtered through their balleen plates.

Kuroshio Current   *noun*
The Kuroshio Current, or Japan Current, is a warm **ocean current.** It flows north, along the Pacific coast of Japan, then turns east to join the **North Pacific Drift**.
In southern Japan, coral reefs flourish in the warm waters of the Kuroshio Current.

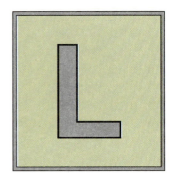

Labrador Current   *noun*
The Labrador Current is a cold **ocean current**. It flows southwards along the west coast of Greenland and Canada until it meets the **Gulf Stream**.
Foggy weather is common where the warm Gulf Stream and the cold Labrador Current meet.

Labrador Sea   *noun*
The Labrador Sea is an arm of the north-west **Atlantic Ocean**. It lies between the south-west coast of Baffin Island and the mainland coasts of Canada and Greenland.
There are rich cod fishing grounds in the Labrador Sea.

lagoon   *noun*
A lagoon is a shallow lake of salt water between the coast and a **barrier island**, sand bar, shingle spit or **coral reef**. Lagoons are connected to the sea at one or more points.
Sea turtles swam into the lagoon to feed on the seagrasses growing there.

lamp shell   *noun*
A lamp shell looks like a **bivalve**, except that the lower shell is usually larger than the upper shell and it has a beak-like tip. Inside the shell is a special filtering structure that looks like a coiled ribbon, fringed with tiny hairs. This filters food **particles** from the water. Some lamp shells have long stalks, while others are attached directly to rocks.
The children found some lamp shells which were anchored in burrows in the sediment.

lamprey   *noun*
A lamprey is a **jawless fish** which lives in the **sea** and in freshwater. Most lampreys are **parasites** of other fish. Their mouth is made up of a round, sucking organ and a toothed tongue. Some lampreys use the sucking organ to attach themselves to other fish. They use their tongue to cut into their **prey** and feed on its blood.
Lampreys breed larvae that live in burrows on river beds.

lanternfish   *noun*
A lanternfish is one of a large group of small **deep sea fish** which have **light organs** along their head, body and tail. There are more than 230 species of lanternfish. Most adult lanternfish measure less than 15 centimetres long. Lanternfish often travel in large shoals consisting of hundreds of thousands of fish. During the day, these shoals are found at great depths, but at night, they migrate to shallower depths.
Most lanternfish live in the deep waters of the open sea in all parts of the world.

larvae ► metamorphosis

lateral line system   *noun*
The lateral line system is a structure in **fish** that picks up vibrations in the water. It is made up of a series of tiny, sensitive hairs inside a groove that runs along the fish's flanks and over its head. Vibrations in the water move tiny hairs, and these send signals to the fish's brain.
The fish's lateral line system picked up the movements of a nearby shrimp.

latitude   *noun*

Latitude is an imaginary line drawn around the **Earth**. It shows exactly where a place is on the Earth's surface. Lines of latitude run parallel to the **Equator**. There are 90 degrees of latitude on each side of the Equator. The Equator is 0 degrees latitude, and the poles are 90 degrees north or south of the Equator.

He plotted the ship's position as latitude 15 degrees south, 165 degrees west.

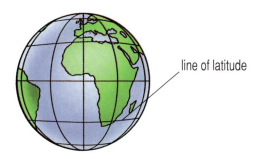

line of latitude

lava   *noun*

Lava is molten rock that flows out onto the Earth's surface from inside the Earth.

Red-hot lava poured from the volcano.

Law of the Sea   *noun*

The Law of the Sea is an agreement between **maritime** countries about who controls various areas in the Earth's seas and oceans. The present Law of the Sea was agreed at a United Nations convention in 1982.

The Law of the Sea states that a country can claim sea within 12 nautical miles of its coast.

lichen   *noun*

A lichen is a small, crusty growth found on rocks, walls and tree trunks. Lichens are made up of a fungus and an alga living together in a form of **symbiosis**. The alga supplies the lichen with food made by **photosynthesis**, and the lichen protects the alga against drying out. Lichens grow very slowly, and can live for long periods without water.

The rocks were covered in yellow lichens.

life raft   *noun*

A life raft is a small inflatable vessel, usually equipped with life-saving gear. Life rafts are designed to stay afloat and inflate quickly. They are used by ships and aeroplanes in emergencies.

Life rafts carry first aid packs and flares.

lifeboat   *noun*

A lifeboat is a boat which is designed to save people from the sea in an emergency. Inshore lifeboats are launched from the shore. They are specially built to travel quickly through rough seas and can right themselves if they capsize. Other lifeboats are simply small boats launched from ships in distress. Lifeboats carry flares. These are coloured rockets that are fired into the sky to show rescuers where to look.

As the ship sank, the captain ordered the crew to launch the lifeboat.

light organ   *noun*

A light organ is a light-producing structure found on the body of fish and other marine animals. The light is produced by chemical reactions, or by bacteria living in the light organs. **Deep sea fish** have light organs which produce light to frighten or confuse **predators**. Light organs are often arranged in patterns. They enable fish belonging to the same species to recognize each other in the dark waters under the ocean.

Flashlight fish use light organs under their eyes to see their prey at night.

73

lighthouse *noun*
A lighthouse is a tall, narrow building with a lamp at the top. The lamp is surrounded by a system of lenses to focus the light into an intense beam. The beam rotates, so the light appears to flash. Its flashing lights shine in the dark or in bad weather to warn sailors of rocks, **reefs** and **islands**.
Each lighthouse has its own special pattern of flashes, so sailors can recognize it.

lightship *noun*
A lightship is a vessel which is moored in a certain position on the sea. It has a light on its mast, a fog signal and radio equipment. Lightships look like floating lighthouses. They are used to mark **shipping lanes** and to warn sailors of rocks.
It was getting dark, but the sailors could see the lightship marking the mouth of the estuary.

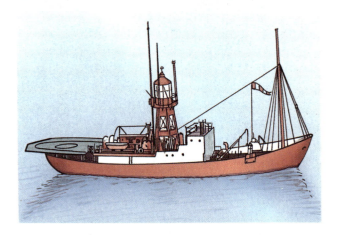

limestone *noun*
Limestone is a **sedimentary rock** made from the shells of marine animals and of single-celled creatures such as **algae** and **foraminiferans**. The shells, which are made mainly of **calcium carbonate**, sink to the **ocean floor** when the animal dies, and become covered in **sediment**. After a long period of time, the shells turn into rock. A limestone called coquina is made up of corals and shells. It is used to make roads.
We found fossils in the limestone cliffs.

limpet *noun*
A limpet is a **gastropod**, with a cone-shaped shell. It lives mainly between the high and low water mark. Limpets cling tightly to rocks with a muscular foot. They move away to graze on **algae** when the tide is in. In most limpets the shell completely covers the animal and protects it from crabs and seabirds. Limpets live on rocky shores in many parts of the world. They are usually less than eight centimetres in length.
Most shorebirds find it impossible to remove a limpet from its rock.

lionfish *noun*
The lionfish is a large **scorpionfish**. It has fins which end in long spines with poisonous tips. It also has bright red and white stripes to warn of danger.
The skin-diver took care to avoid the long spines of the lionfish.

littoral zone ► **intertidal zone**

lizardfish *noun*
A lizardfish is a small **fish** that shuffles across the sea-bed, using its pelvic **fins** like legs. When it is propped up on its fins, waiting for **prey** to pass by, it looks rather like a lizard.
As a lizardfish darts out to catch its prey, it spreads its pectoral fins like wings.

lobster *noun*
A lobster is a large marine **crustacean**. It has long feelers or antennae, four pairs of walking legs, which are also used for swimming and a pair of large pincers. It has a long body, which ends in a fan-like tail. Most kinds of lobster have dark green or dark blue shells with spots on them. When they are cooked, the shell turns bright red. Lobsters live on the sea-bed. They are found in the Atlantic and Pacific Oceans. Lobsters come out to hunt and scavenge at night.
A large lobster was lurking in the rock crevice.

lobster pot *noun*
A lobster pot is a wooden or plastic trap for catching **lobsters**. Attracted by bait inside the trap, the lobster goes in through a tunnel-like opening made of net, which allows it to enter, but not leave.
They collected lobsters from the lobster pots.

long-line fishing *noun*
Long-line fishing is a way of catching fish such as cod, which live near the sea-bed. A line of baited hooks is laid along the sea-bed with an **anchor** at each end attached to a **buoy** at the surface.
The fishermen decided to use the long-line fishing method instead of trawling.
long-line fishing *verb*

long-shore drift *noun*
Long-shore drift describes the way in which **sand** or shingle is moved sideways along a beach by the force of tidal currents. Where the waves approach the beach at an angle, they carry material diagonally up the beach. As they retreat, they drag material directly back into the sea by the shortest route.
Sand has piled up against the windbreak as a result of the long-shore drift.

longitude *noun*
Longitude is an imaginary line around the **Earth**. It begins at the North Pole and ends at the South Pole. Longitude is a measurement used for distance eastwards or westwards.
There are 360 degrees of longitude around the Earth.

loosejaw *noun*
A loosejaw is a small, slender **deep sea fish** with dark skin and small, scattered **light organs**. Its mouth has no floor, and it has very long jaws armed with sharp needle-like teeth. The mouth can be opened very wide.
Unlike most fish, loosejaws can throw their heads back to capture large prey.

low pressure area ▶ **depression**

low water mark *noun*
The low water mark is the lowest point on a beach that remains uncovered at low water on a **spring tide**.
Animals that live above the low water mark must be able to cope with pounding by waves.

lugworm *noun*
A lugworm, is a fleshy **worm**, which lives in burrows found in muddy **sand**, in the **intertidal zone**. It has bushy **gills** on the front part of its body, and draws water through its burrow for breathing. The lugworm takes in sand and **mud** as it burrows, and extracts **detritus** from it. Undigested sand and mud pass out of the worm and form coiled 'casts' on the **shore**.
The fishermen were digging for lugworms to use as bait.

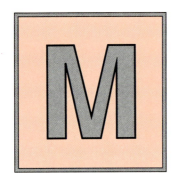

mackerels and tunas ► page 80

maelstrom *noun*
A maelstrom is a powerful and dangerous **current** found in the **Arctic Ocean**. It sweeps backwards and forwards between two islands off the western coast of Norway. The maelstrom becomes more dangerous when the wind blows against it, between high and low tide.
The waters of the maelstrom form huge whirlpools which can destroy small ships.

magma *noun*
Magma is red-hot liquid, or molten, rock lying beneath the Earth's surface. Once magma reaches the Earth's surface, it is called lava. Magma may cool to form solid rocks such as granite.
The magma formed a stream of red-hot lava.

magnet *noun*
A magnet is an object, usually made of metal, which attracts other metal objects. A magnet has two magnetic poles, 'north' and 'south'. The north pole of the magnet attracts only the south poles of other magnet, and vice versa.
The Earth's core contains a large amount of metal and acts as a giant magnet.
magnetic *adjective*

Malacca Strait *noun*
The Malacca Strait is a channel between Sumatra and the Malaysian **peninsula**. It links the **Indian** and **Pacific Oceans**.
The Malacca Strait is one of the world's busiest shipping lanes.

mammal *noun*
A mammal is a warm-blooded **vertebrate** that suckles its young with milk. Many mammals are covered in fur. Most marine mammals do not have fur. Instead, they have a layer of **blubber** which keeps them warm. **Whales, seals, walruses** and **sea otters** are all mammals.
Marine mammals cannot stay under water for very long, as they have to breathe air.

manatee ► **sirenian**

manganese nodule *noun*
A manganese nodule is a small, brown, potato-shaped lump which contains the metal manganese. Manganese nodules cover large areas of the deep ocean floor.
The manganese nodules on the floor of the Pacific Ocean contain about 40 billion tonnes of manganese.

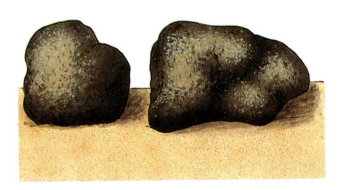

Mangrove swamp *noun*
A mangrove swamp is a marshy part of tropical and sub-tropical **coasts**. It is covered in mangroves. Mangroves are shrubs and trees which have adapted to being flooded by the salty sea. Most of them have shiny, evergreen leaves, and roots that arch down from the trunk into the mud. Some also have breathing roots which stick up out of the mud. Mangrove swamps are also known as mangrove forests. They can sometimes be several kilometres wide. Fish such as **milkfish** and **groupers** breed in mangrove swamps.
Fiddler crabs often live in mangrove swamps.

manta ray   *noun*
A manta ray, or devil ray, is the largest of the ray family. Some may be over six metres across. Manta rays can weigh more than a tonne. They live in the open **sea**. Manta rays have two **fins** on each side of the head. These roll up like horns when swimming, which has led to the name 'devil ray'. When feeding, these fins are spread out to direct **shoals** of **fish** or **shrimps** into the manta's wide mouth.
The manta ray had driven a shoal of fish into the shallow water to force them into its mouth.

mantis shrimp   *noun*
A mantis shrimp, or stomatopod, has huge folded legs like those of praying mantis. It lives in burrows in the sea-bed, where it lies in wait for **prey**. When a small **crustacean** comes within reach, these legs snap out and deliver a stunning blow. Mantis shrimps also use their weapons to fight each other for territory on the sea-bed.
The prawn was stunned by a blow from the mantis shrimp.

mantle   *noun*
The mantle is a thick layer of solid rock which lies between the Earth's crust and the core. It is mainly solid rock, but some of the rocks are molten. The rock in the mantle is made of oxygen, silicon, iron, aluminium and magnesium.
The movement of rocks in the mantle cause continental drift and sea-floor spreading.

Marianas Trench   *noun*
The Marianas Trench is a deep trench in the **ocean floor** in the west **Pacific Ocean**. This is the deepest known part of the ocean.
In parts of the Marianas Trench, the water is over 11,000 metres deep.

mariculture ▶ **aquaculture**

marina   *noun*
A marina is an area beside the **sea**, or in a dock or artificial **harbour**, where motorboats and yachts are moored.
She kept his new yacht moored at the marina.

marine   *adjective*
Marine describes something that is found in, or comes from the sea.
Many marine mammals have flippers instead of legs.

marine archaeology   *noun*
Marine archaeology is the study of the remains of ancient civilizations that now lie under the surface of the sea. It includes the study of wrecked ships and drowned cities.
She wanted to study marine archaeology because she enjoyed history.

marine biologist   *noun*
A marine biologist is a person who studies or practises **marine biology**.
He is working as a marine biologist for the Fisheries Protection Service.

marine biology   *noun*
Marine biology is the study of plants, animals and other organisms that live in the sea.
They set sail on the research ship to study the marine biology of the Pacific Ocean.

marine geologist   *noun*
A marine geologist is a person who studies or practises **marine geology**.
A team of marine geologists was drilling a hole in the sea-bed.

mackerels and tunas *noun*

Mackerels and tunas are smooth, silvery fish that are found in the open **ocean**. They have a streamlined, torpedo-shaped body with a deeply-forked tail that becomes very narrow where it joins the body. In front of the tail are strings of small finlets. Mackerels and tunas live in large **shoals** and swim very fast. *Fishing boats follow migrating shoals of mackerels and tunas.*

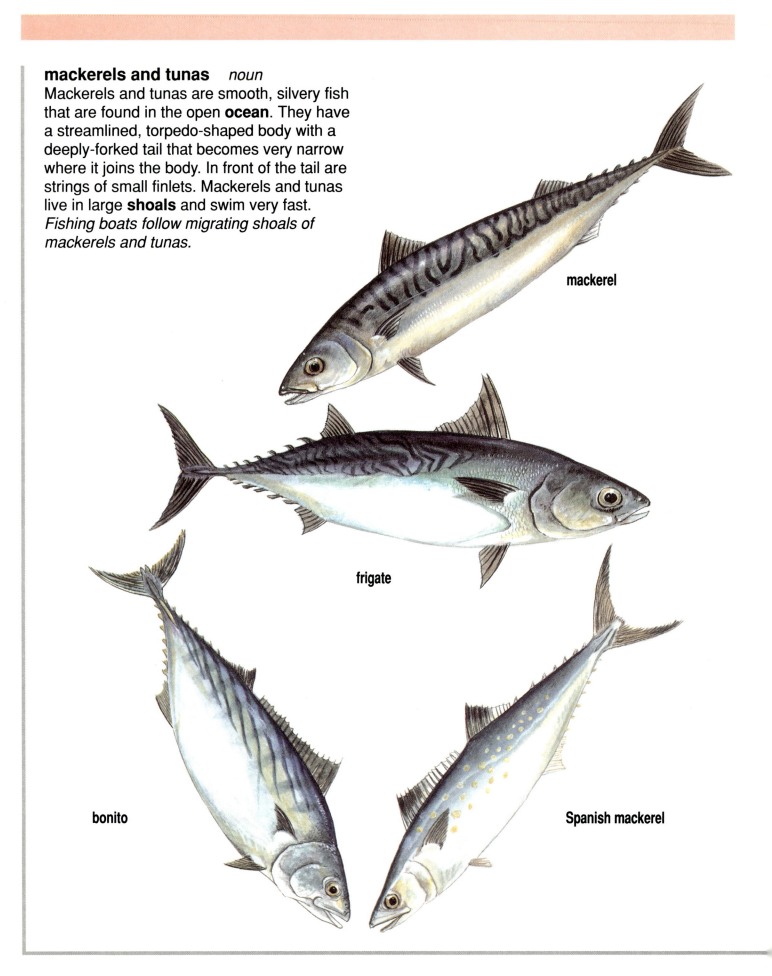

mackerel

frigate

bonito

Spanish mackerel

blue fin tuna

little tunny

albacore

yellow fin tuna

marine geology *noun*
Marine geology is the study of the rocks and **sediments** of the **ocean floor** and the processes that formed them.
She was studying marine geology at the university.

marine iguana *noun*
The marine iguana is the world's only marine lizard. It is large, with a blunt snout and a spiny crest on its neck. It has webbed feet for swimming and long claws for clinging to rocks in the surf. Marine iguanas are found only on the Galapagos Islands, off the coast of Ecuador in South America. They feed on **algae** near the shore.
The marine iguanas were sunbathing on the rocks by the sea.

marine snow *noun*
Marine snow is the name given to tiny particles in sea water that have been formed from the bodies of dead or living organisms. It includes animal droppings and cast-off shells.
Many deep sea animals live on marine snow.

maritime *adjective*
Maritime describes something which is near the sea or which is affected by the sea. For example, a **maritime climate** has mild weather all year, and there is no dry season.
The maritime countries of the world have agreed to obey the Law of the Sea.

maritime climate *noun*
A maritime climate is a climate which has mild weather all year and no dry season. Rain can fall in any season in this climate.
The maritime climate of south-west England is good for dairy farming.

marlin *noun*
A marlin is a large fish with upper jaws that extend into a long spike. It uses the spike to stun small fish. The marlin's tail is deeply forked.
The sport fishermen caught a large marlin.

maroon *noun*
A maroon is a noisy firework, or flare, that is used to raise the alarm if a boat is in trouble at **sea**.
They heard the maroons, and rushed to the beach to watch the lifeboat being launched.

marram grass *noun*
Marram grass is a tough wiry grass that grows on **sand dunes**. It can cope with being buried by sand. Its stiff clumps help to trap the sand and stop the dunes being blown away or moving inland.
They planted marram grass on the dunes to stabilize them.

Maury, Matthew Fontaine (1806-1873)
Matthew Fontaine Maury was a scientist who published the first textbook of modern oceanography.
Matthew Fontaine Maury compiled modern charts of the oceans and maps of wind belts.

Mediterranean Sea *noun*
The Mediterranean Sea is a large inland sea. It is surrounded by Europe to the west and north, Asia to the east, and Africa to the south. The Mediterranean Sea is linked to the **Atlantic Ocean** by the Strait of Gibraltar, to the **Black Sea** through the **Sea of Marmara**, and to the **Red Sea** by the **Suez Canal**.
The coasts surrounding the Mediterranean Sea are very popular with holiday makers.

medusa *noun*
Medusa is the name biologists use to describe a **jellyfish**.
Some medusa have very painful stings.

menhaden ► **herring family**

merchant marine *noun*
The merchant marine, or merchant navy is the name given to a fleet of commercial, passenger and cargo ships. It does not include naval vessels or pleasure craft. Merchant ships carry goods and people from place to place.
Commercial fishing vessels do not belong to the merchant marine.

merchant seaman ► **merchant ship**

merchant ship *noun*
A merchant ship is a ship that carries goods rather than people. A country's merchant fleet is made up of ships involved in commercial activities rather than military activities.
A sailor on a merchant ship is called a merchant seaman.

mermaid *noun*
A mermaid, or siren, was a mythical creature that lived in the sea. Mermaids were thought to be half fish, half human. Sailors used to think the **sea cow** was a mermaid, which is why this animal was also called **sirenian**.
In folklore, mermaids had magical powers.

mermaid's purse *noun*
A mermaid's purse is the egg case of a **skate**.
The curling corners of a mermaid's purse prevent it from floating out to sea.

metamorphosis *noun*
Metamorphosis describes the way some kinds of animal change shape as they grow from young to adult. The young look quite different from their parents and are called **larvae**. Many marine animals go through several stages of metamorphosis before they become adults. Such animals include **molluscs, echinoderms** and **crustaceans**.
Young barnacles pass through several stages of metamorphosis in the plankton before they settle out on the sea-bed.
metamorphose *verb*

microscopic *adjective*
Microscopic describes an object that is so small that it can be seen only with a microscope.
Baby herrings feed on the microscopic creatures of plankton.

Mid-Atlantic Ridge *noun*
The Mid-Atlantic Ridge is a huge underwater mountain chain running from north to south through the **Atlantic Ocean**. It is part of the **mid-ocean ridge system**. In places, the mountains rise above the water as islands.
The island of Iceland lies on the Mid-Atlantic Ridge.

mid-ocean ridge system   *noun*
The mid-ocean ridge system is a gigantic underwater mountain chain that branches through all the world's oceans. In some places it is 5,000 kilometres wide. Sometimes the mountains rise above the water to form islands. Along these ridges, the **tectonic plates** carrying the **ocean floor** are moving apart.
Lava wells up along the mid-oceanic ridge system to form mountains and new sea floor.

migration   *noun*
Migration is the regular movement of animals between two or more places. Many animals use special places for breeding, or move to more sheltered waters in winter. **Whales** migrate from the plankton-rich polar waters to warm **lagoons** to give birth. **Sea turtles** migrate to sandy **beaches** to lay their eggs. European **eels** migrate from rivers to the **Sargasso Sea** in order to **spawn**.
The tourists travelled to the west coast of the United States of America to watch the migration of the whales.
migrate   *verb*

milkfish   *noun*
The milkfish is a very important food fish. It is found in the warmer parts of the **Pacific** and **Indian Oceans**. It is a large, silvery fish with a deeply forked tail. The milkfish migrate to shallow coastal waters to **spawn**. They can reach a length of approximately 120 centimetres and can weigh about 13 kilogrammes.
In south-east Asia, people collect young milkfish to rear by aquaculture.

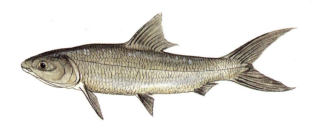

milt   *noun*
Milt is the fluid shed by a male fish during mating, or **spawning**. It contains the male's sex cells, which join with the female's eggs to produce a new fish.
The female grunion burrows into the sand, and the male then sheds its milt on the eggs.

mining   *noun*
Mining is the process of taking metals, valuable stones and minerals from the Earth. Mining under the sea involves the extraction of nodules, or lumps of minerals, which are found on the sea-bed. These nodules usually contain iron and manganese, which are used to make hard steel for tools.
Mining for manganese nodules is still an expensive process.
mine   *verb*

Moho ► **Mohorovicic discontinuity**

Mohorovicic discontinuity   *noun*
The Mohorovicic discontinuity, or Moho, is the boundary between the Earth's crust and the mantle.
The rocks beneath the Mohorovicic discontinuity are much denser than the rocks above.

mollusc   *noun*
A mollusc is an **invertebrate**. It has a soft body which is often protected by a hard shell. Molluscs use **gills** to breathe. Most molluscs feed using a ribbon-like tongue, called a radula, which is armed with horny teeth. Some molluscs, like cuttlefish, have no outside shell. Instead, they have a special shell that grows inside their body. In cuttlefish this is called a **cuttlebone**. Some molluscs, like the octopus, have no shell at all. The main groups of molluscs are **chitons**, **bivalves**, **gastropods**, **cephalopods** and tusk shells.
There are over 100,000 kinds of mollusc.

monkfish ► **shark**

moonfish *noun*
A moonfish, or opah, is a large **fish** with a thin, flattened body. It has beautiful colours with a steel blue back, silvery sides and a green abdomen tinged with rose pink. The moonfish is covered in silver spots. Its fins, tail and jaws are bright red. It is a large fish, measuring up to two metres long, and can weigh over 140 kilogrammes.
Fishermen in warm oceans may occasionally catch a moonfish, which is very good to eat.

Moorish idol *noun*
The Moorish idol is a kind of **surgeonfish** with a very flattened body and bold, broad vertical stripes of black and yellow edged with creamy white. It has a long anal **fin** and a tall dorsal fin that tapers to a curving point, a narrow snout with a bold orange patch, and horn-like knobs over its eyes.
The Moorish idol's bold pattern helps to disguise its shape against the bright colours of the coral reef.

moss animal ► **sea mat**

mother-of-pearl *noun*
Mother-of-pearl is a kind of calcium carbonate that lines the shells of some kinds of **bivalve**. It is rainbow-coloured. Shells containing mother-of-pearl are found off the coasts of tropical countries such as Australia and the Philippine islands.
She wore a mother-of-pearl necklace.

Mozambique Channel *noun*
The Mozambique Channel is a **strait**, found between Madagascar and the African mainland.
The Mozambique Channel is a busy shipping route.

mucus *noun*
Mucus is a slimy substance produced by some animals. Mucus retains water and also makes surfaces slippery. Fish are covered in mucus to help them slip through the water.
Seaweeds are covered in mucus that helps to prevent them drying out at low tide.

mud *noun*
Mud is a sediment made up of particles that are smaller than **sand**. These are usually no more than a fraction of a millimetre across.
The river dropped its mud as it entered the calm waters of the estuary, building up mudflats.

mudflat *noun*
A mudflat is a large, flat area of **mud**.
Many small birds were feeding on the mudflats on either side of the estuary.

mudskipper *noun*
A mudskipper is a **goby** that lives on **mudflats** and in **mangrove swamps**. It can survive for some time out of water. Mudskippers store air and water in their gill chambers and also take in oxygen from the air through the lining of their mouth and throat. They have thick, fleshy pelvic fins which they use like stubby legs to hop over the mud and climb mangrove roots.
A mudskipper sitting on a mangrove root was trying to catch a passing fly.

mullet *noun*
A mullet is a silvery fish with large scales.
Mullets live in large shoals and are caught
for food. They feed on **invertebrates** and
algae in shallow water.
*The red mullet uses long barbels on its chin
to feel in the mud for its prey.*

murre ▶ **auk**

nansen bottle *noun*
A nansen bottle is a container shaped like a
cylinder, which can be lowered over the side
of a ship. Once it reaches its destination, it
can be sealed to allow a sample of sea
water to be taken. Back on board the ship,
the nansen bottle is opened and the sample
of sea water can be analysed.
*He opened the nansen bottle to test the
water sample inside.*

narwhal *noun*
A narwhal is a small Arctic **whale** that feeds
on fish, **cephalopods** and **crustaceans**. It
has only two teeth which are at the front of
the jaws. In males, one tooth forms a spirally
twisted tusk which can be up to 2.7 metres
long. Narwhals live in the Arctic Ocean and
eat mainly shrimp, fish, squid and **octopus**.
*No one knows what the narwhal uses its
long tusk for.*

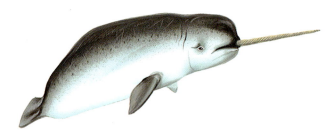

nautical mile *noun*
A nautical mile is a measure of distance at
sea. One international nautical mile is equal
to 1.85 kilometres. One **knot** is equal to one
nautical mile an hour.
*The merchant ship was 100 nautical miles
from the coast of Africa.*

nautilus *noun*

A nautilus is a **cephalopod** with a smooth, coiled shell. The shell of the nautilus is divided into chambers. The nautilus lives in the outermost chamber. It swims near the **ocean floor** and feeds on **crustaceans**.
By adjusting the amount of gas in its chambers, the nautilus can rise or sink in the water.

naval ships ► page 86

navigation *noun*

1. Navigation is a science which involves plotting a course around the world's **oceans**. A ship's navigator has charts or maps of the ocean, showing coastlines and **depth** of water. The ship's **chronometer** and **sextant**, along with the **compass**, allows the ship's position to be calculated. Today, most ships use satellite navigation, or Satnav. Information from satellites orbiting the Earth, is received by computer on board the ship. Two sets of figures give the **longitude** and **latitude** of the vessel to within a few metres.
2. Navigation describes how an animal finds its way around. Some animals, such as **seals** and **whales**, may use landmarks along the coast to help them. Others rely on their different senses. Eels, for example, use taste and smell, and **dolphins** use **echo-location**. **Sandhoppers** are guided by the positions of the Sun and the Moon or by tiny variations in the pull of **gravity**, to find their way up and down the **beach**.
Green turtles use ocean currents for navigation.
navigate *verb*

neap tide *noun*

The neap tide is the tide with the smallest difference in water level between **high tide** and **low tide**. Neap tides take place near the time of the Moon's quarters, when the gravities of the Sun and the Moon pull in different directions. The opposite of a neap tide is a **spring tide**.
The rocks were exposed during neap tides.

needlefish *noun*

A needlefish is a long, thin **fish** with a narrow, needle-like snout. When it is disturbed, it leaps out of the water and sails through the air like an arrow. At night, arrowfish sometimes leap towards lights, and have been known to badly injure fishermen who get in their way.
A needlefish jumped across the surface of the water in front of the boat.

nekton *noun*

Nekton describes the group of animals which swim freely in the open ocean. These animals do not drift with currents or **tides**, but can move in any direction.
Fish make up a large part of the nekton.

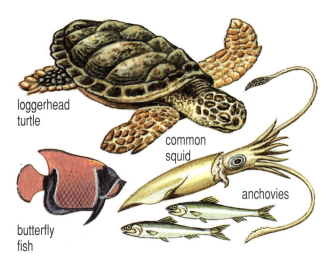

loggerhead turtle

common squid

anchovies

butterfly fish

nematocyst *noun*

Nematocyst is the name given to a stinging cell found on the **tentacles** of **sea anemones**, hydroids, **jellyfish** and other **cnidarians**.
As the fish touched the jellyfish's tentacles, poison shot through the nematocysts and stung it.

Neptune *noun*

Neptune was the Roman god of the sea. He was worshipped in the same way as the Greek God Poseidon, who was also God of the sea.
The Lateran Museum in Rome Italy, has a splendid statue of Neptune holding a trident.

naval vessel *noun*

A naval vessel is a ship or boat used by the navy. Naval vessels often include warships which are designed to take part in war. **Frigates**, **submarines**, aircraft carriers and destroyers are all kinds of naval vessel.
The navy sent out various naval vessels to sea.

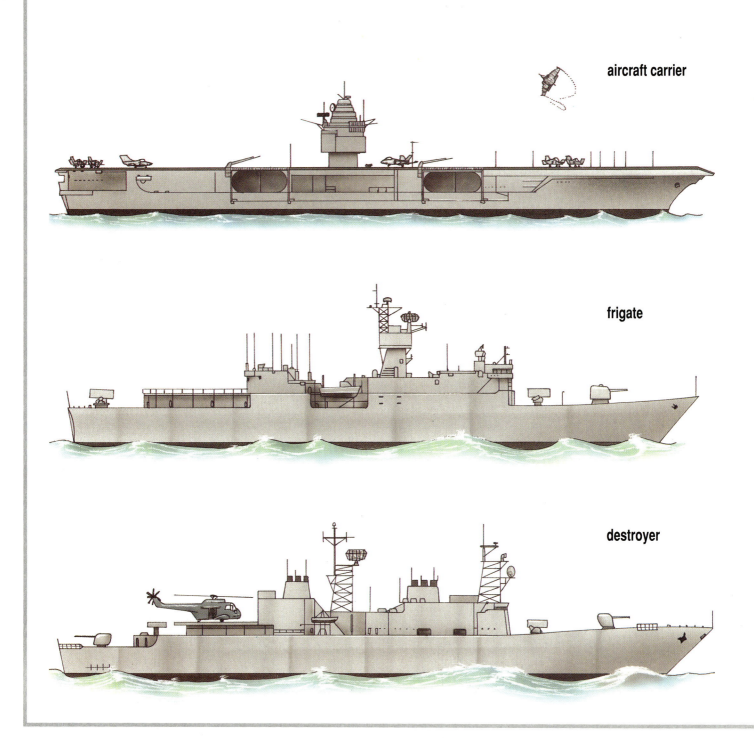

aircraft carrier

frigate

destroyer

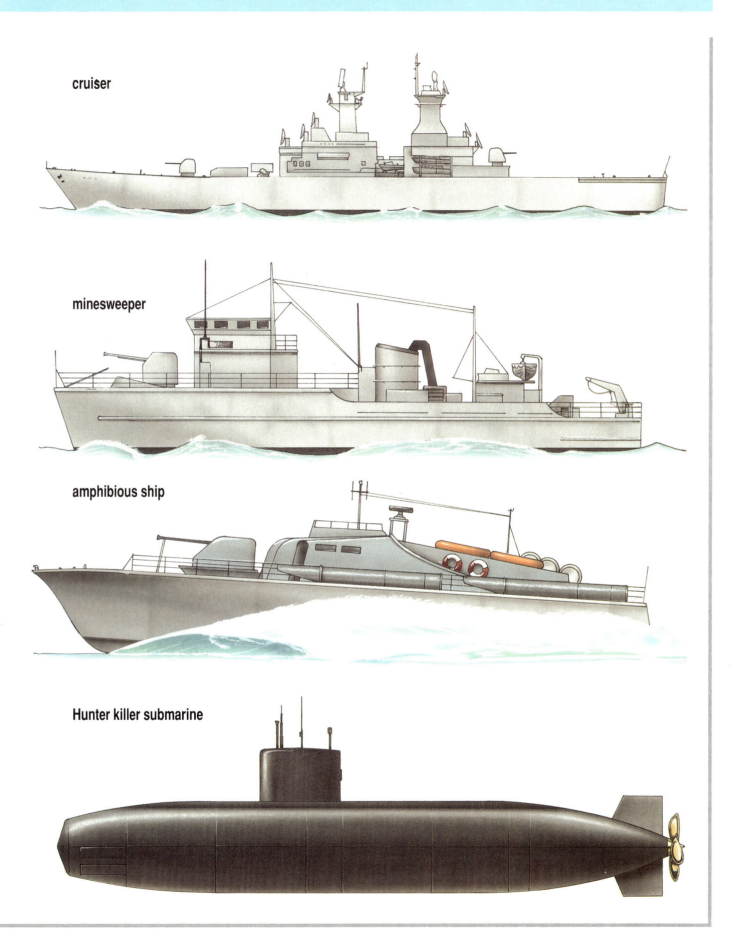

cruiser

minesweeper

amphibious ship

Hunter killer submarine

net *noun*
Net is a material made of twine, cord or synthetic fibres woven together to form a set of meshes. Nets are used to trap fish and other animals. The size of the spaces in the mesh depends on the size of the animals to be caught. Different types of net include **purse seine nets** and trawl nets. Some fishing nets can weigh 4,500 kilograms and are as long as 600 metres.
The fishermen were mending their nets.
net *verb*

nitrogen *noun*
Nitrogen is a gas. It makes up 78 per cent of the Earth's atmosphere. The element nitrogen is an important part of proteins and other essential chemicals in the body.
Nitrogen is present in sea water as a part of chemicals called nitrates.

non-tidal current *noun*
A non-tidal current is a current that is not caused by tidal forces. **Surface currents** which are caused by the wind and currents that are due to the shape of the ocean floor, are non-tidal currents.
Non-tidal currents sweep sediment down the continental slope.

North Atlantic Drift *noun*
The North Atlantic Drift, or North Atlantic Current, is an **ocean current**. It flows north-east, across the **Atlantic Ocean** from the coast of Newfoundland in Canada, to the **Norwegian Sea**.
Warm water from the Gulf Stream flows into the North Atlantic Drift.

North Equatorial Current ► **Equatorial Current**

North Pacific Drift *noun*
The North Pacific Drift, or North Pacific Current, is a warm **ocean current** that flows westwards across the North **Pacific Ocean**.
The North Pacific Drift is part of the North Pacific gyre.

North Sea *noun*
The North Sea is an arm of the North Atlantic Ocean. It lies between the continent of Europe to the south and east, and the United Kingdom to the west.
There are important fisheries as well as oil and gas fields in the North Sea.

North-west Passage *noun*
The North-west Passage is the sea route found along the north coast of North America, between the Atlantic and Pacific Oceans.
Many oil tankers use the North-west Passage.

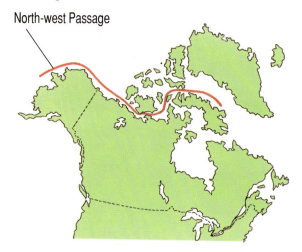
North-west Passage

Norway Current *noun*
The Norway Current is a branch of the warm **North Atlantic Drift** which flows along the coast of Norway into the **Barents Sea**.
The warmth of the Norway Current helps prevent Norwegian ports from freezing in winter.

Norwegian Sea *noun*
The Norwegian Sea is part of the North **Atlantic Ocean**. It is found between Greenland and Iceland in the west and Norway in the east.
Fishermen from many countries fish for cod and herring in the Norwegian Sea.

nuclear waste ► **radioactive waste**

nudibranch ► **sea slug**

nutrient *noun*
A nutrient is a substance that living organisms need in order to live and grow. Proteins, fats, carbohydrates and minerals are all nutrients.
The nutrients in the sewage encouraged the growth of huge blooms of poisonous algae.

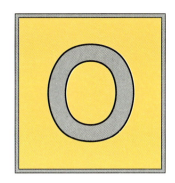

oarfish *noun*
An oarfish is a ribbon-like **fish**. It can measure up to 11 metres long and about 30 centimetres deep. The oarfish is a silvery blue transparent colour, with dark diagonal stripes and a bright red **fin**, which runs along the entire length of its body. It swims with a snake-like motion. The oarfish is also known as the king-of-the-herrings.
The oarfish is so large that it may be responsible for many myths about sea monsters.

ocean ► page 90

ocean basin *noun*
The ocean basin is that part of the ocean beyond the edges of the **continental shelves**. The **continental slopes** form the edge of the ocean basin.
The ocean basin contains flat plains, submarine mountains, long, narrow valleys and deep trenches.

ocean circulation *noun*
Ocean circulation describes the movement of the water on the surface of the planet. There are water currents at all depths of the ocean. Surface currents are caused by prevailing winds. Deep water currents are caused by differences in **salinity** and temperature. Ocean circulation is affected by the rotation of the Earth and the effects of the Sun's radiation.
The pattern of the ocean circulation affects the world's climate.

ocean current ► page 92

ocean *noun*

An ocean is a very large sea which usually separates continents. There are five oceans. These are, in order of size, the Pacific, Atlantic, Indian and Arctic oceans. Together, these cover 71 per cent of the Earth's surface. **Seas**, **bays** and **gulfs** are smaller bodies of water that lie between islands and land masses.

The passenger ship crossed the Atlantic Ocean.

ARCTIC OCEAN

Baffin Bay

Davis Strait

Bering Sea

Hudson Bay

NORTH PACIFIC OCEAN

NORTH ATLA OCEAN

Gulf of Mexico

Gulf of California

Caribbean Sea

SOUTH PACIFIC OCEAN

SC

Weddel Se

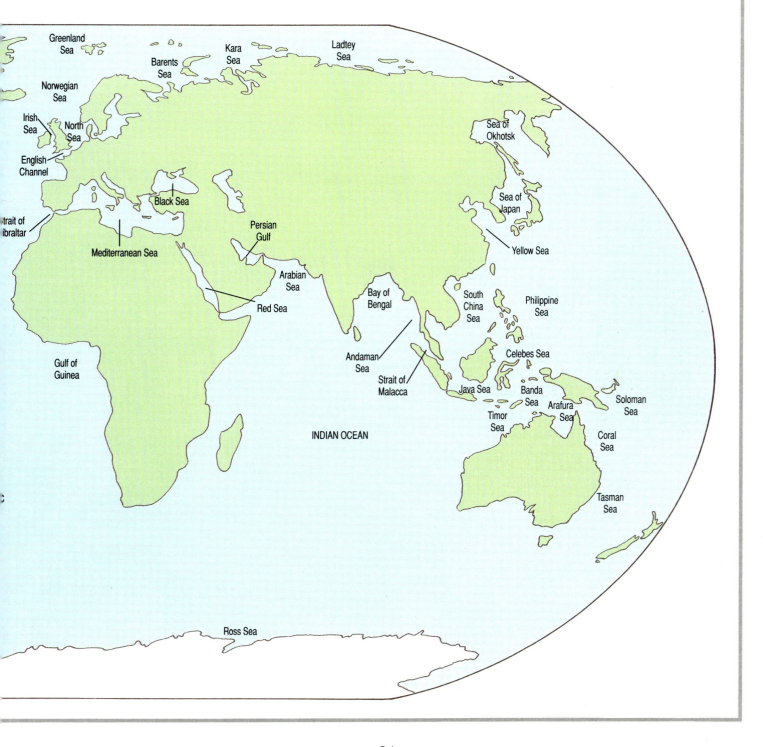

Greenland
Sea

Kara
Sea

Ladtey
Sea

Barents
Sea

Norwegian
Sea

Sea of
Okhotsk

Irish
Sea

North
Sea

English
Channel

Sea of
Japan

Black Sea

trait of
ibraltar

Persian
Gulf

Yellow Sea

Mediterranean Sea

Arabian
Sea

South
China
Sea

Philippine
Sea

Bay of
Bengal

Red Sea

Celebes Sea

Gulf of
Guinea

Andaman
Sea

Strait of
Malacca

Java Sea

Banda
Sea

Soloman
Sea

Arafura
Sea

Timor
Sea

Coral
Sea

INDIAN OCEAN

Tasman
Sea

Ross Sea

ocean current   *noun*

An ocean current is a movement of water in the sea. There are several kinds of ocean current. Tidal currents are related to the movement of the **tides**. Surface currents are caused by the prevailing wind. Differences in density cause currents because dense water flows towards less dense water. Cold water is denser than warm water and salt water is denser than fresh water.

The direction of currents is affected by the shape of the ocean floor.

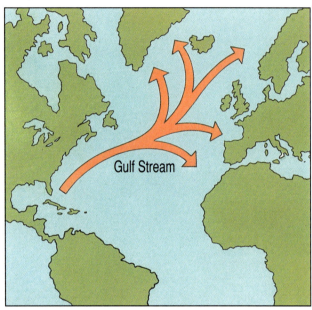

The Gulf Stream carries warm waters from the Gulf of Mexico, north, along the east coast of the United States of America and across the Atlantic Ocean to western Europe. Here, mild winters occur as a result.

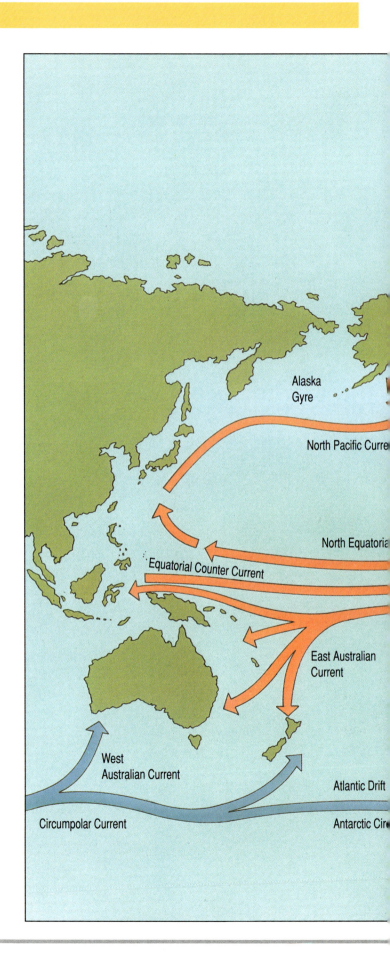

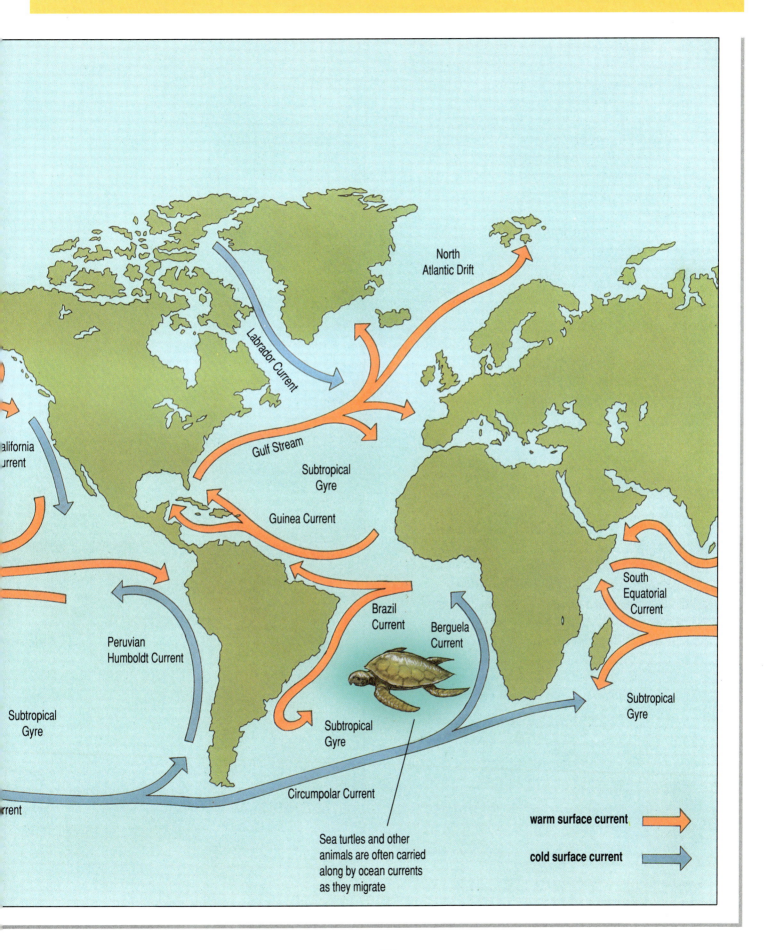

North
Atlantic Drift

Labrador Current

California
Current

Gulf Stream

Subtropical
Gyre

Guinea Current

South
Equatorial
Current

Brazil
Current

Berguela
Current

Peruvian
Humboldt Current

Subtropical
Gyre

Subtropical
Gyre

Subtropical
Gyre

Circumpolar Current

Sea turtles and other
animals are often carried
along by ocean currents
as they migrate

warm surface current

cold surface current

ocean engineer   *noun*
An ocean engineer designs and makes special equipment and tools. These are used to explore the oceans and develop resources such as oil, gas and minerals in the oceans. Ocean engineers may also be involved in building **sea defences**, using the energy from **waves** and **tides**.
Ocean engineers are designing a new kind of submersible which will be used for exploring ocean trenches.

ocean floor ► page 96

ocean liner   *noun*
An ocean liner is a ship that is used to carry passengers and goods, or freight, across the ocean. These trips are normally scheduled, or made at certain times.
The passengers boarded the ocean liner to cross the Atlantic Ocean.

ocean pollution ► **pollution**

ocean strider   *noun*
An ocean strider is an insect that lives on the **ocean** surface. It is the only known **marine** insect. Its feet are designed not to break through the surface film of the water. Ocean striders row themselves across the water with their middle pair of legs, using the hind legs to steer. The front legs are kept free for seizing food, which is mostly the drifting bodies of dead marine animals.
On the purple shell of the bubble raft snail were eggs that an ocean strider had just laid.

ocean zones ► page 98

oceanarium (plural **oceanaria**)   *noun*
An oceanarium is a saltwater **aquarium** for keeping and displaying marine animals and **seaweeds**. Oceanaria are created to interest tourists. They are also used for research in marine biology.
The children visited the oceanarium to study the seals.

oceanographer   *noun*
An oceanographer is a person who studies everything concerned with the oceans. This covers the study of **marine biology**, **marine geology**, chemistry and physics.
The oceanographers said that ocean currents would carry the oil spill away from the shore.

oceanography   *noun*
Oceanography is the study of the oceans.
While studying oceanography, the student learned a great deal of chemistry, geology and biology.

octopus (plural **octopuses**)   *noun*
An octopus is a **cephalopod**. It has a bag-like body and eight tentacles which are covered in suckers. Octopuses have very good eyesight, and a horny beak for feeding on shellfish.
The octopus seized the shrimp in its tentacles.

offshore   *adjective*
Offshore describes something which moves away from the shore, or which is some distance away from the shore. The opposite of offshore is **onshore**.
The offshore breeze blew the balloon out to sea.

oil ► **petroleum**

oil field *noun*
An oil field is the name given to a large underground deposit of crude oil or **petroleum**. Oil fields are found in sedimentary rocks often thousands of metres below the surface of the Earth.
There are many oil fields near the shores of the Gulf of Mexico.

oil platform ▶ page 100

onshore *adjective*
Onshore describes something which moves towards the shore. An onshore breeze is a breeze blowing towards the sea-shore from the sea. The opposite of onshore is **offshore**.
It was too cold to sunbathe on the beach in the fresh, onshore breeze.

ooze *noun*
Ooze is the name given to mud found on the **ocean floor** in deep waters. It contains the remains of **plankton**, which make up at least 30 per cent of its weight.
There were millions of foraminiferan shells in the ooze on the deep sea floor.

opah ▶ **moonfish**

operculum *noun*
An operculum is a flap covering the **gills** of most fish. It is also the horny cover that some **gastropods** use to close their shell opening to protect themselves against enemies.
The cleanerfish wriggled under the jack's operculum to clean its gill chamber.

opisthobranch ▶ **sea hare**

orca ▶ **killer whale**

organism *noun*
An organism is any living thing. The word is used to describe plants, **algae**, animals, **bacteria** and other forms of life.
Sea fans feed on microscopic organisms that drift past in the water currents.

ostracod *noun*
An ostracod, or seed shrimp, is a small **crustacean** found in fresh and **salt water**. It has a bean-shaped shell, which is in two parts. Two pairs of large hairy feelers or antennae, and two pairs of legs protrude through the shells. Most ostracods live near the sea bed and feed on **detritus** and small animals.
Ostracods use their legs and antennae to glide over seaweeds in rock pools.

otter *noun*
An otter is a mammal with a sleek, streamlined body, a fairly long tail, and webbed feet for swimming. Otters are very good swimmers and can stay under water for three to four minutes. They live on all the continents except Australia and Antarctica. **Sea otters** live in salt water, other kinds of otter live mainly in fresh water. Most otters weigh from 4.5 to 15 kilograms. They eat crayfish, **crabs** and fish.
After the oil tanker went aground, we rescued many otters covered in oil.

overfishing *noun*
Overfishing describes the taking of too many fish or other marine animals from the sea. The result of this is that the fish populations become smaller, or decline. If fish or other animals are caught faster than they can reproduce, their numbers will fall.
Whales have suffered badly from overfishing because they reproduce very slowly.

ocean floor   *noun*

The ocean floor, or sea-bed, is the solid base of the **ocean**. It may be covered in **sand** or **mud**, or it may be bare rock. The shape of the ocean floor is as varied as the shape of the dry land.
So much mud and sand collects on the ocean floor, that the weight turns it into rock.

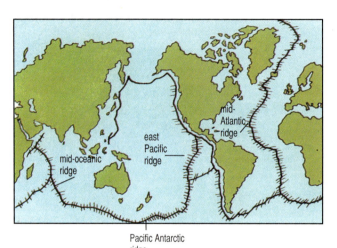

The major oceanic ridges

Coral reefs often form around volcanic islands. If the island sinks back into the sea, the coral reef can carry on growing. The coral ring that is left is called an atoll.

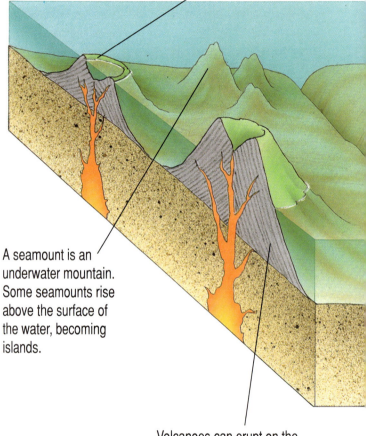

A seamount is an underwater mountain. Some seamounts rise above the surface of the water, becoming islands.

Volcanoes can erupt on the ocean floor. They sometimes erupt above sea-level, creating volcanic islands.

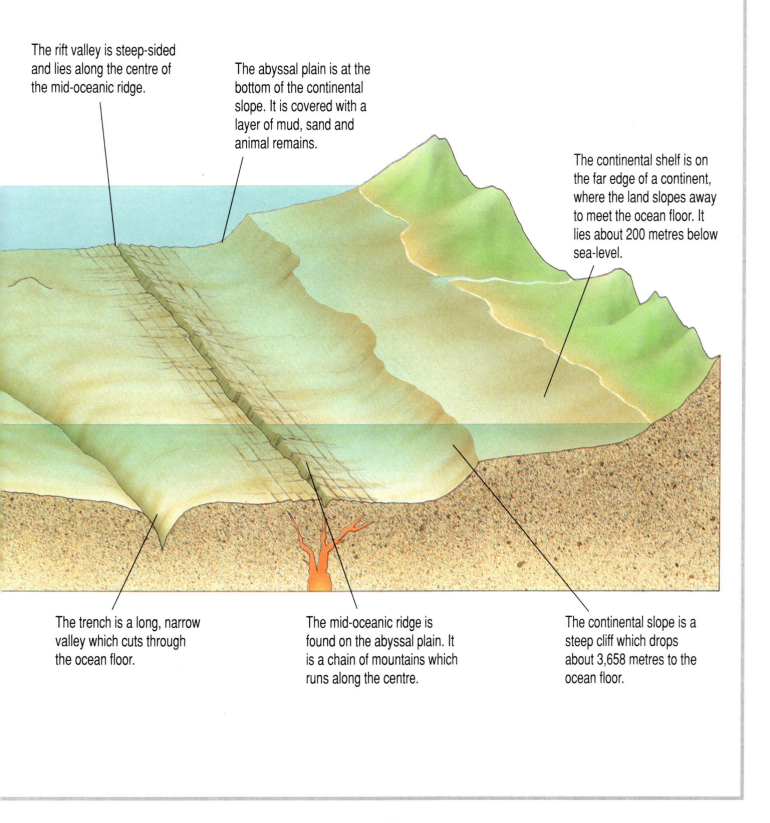

The rift valley is steep-sided and lies along the centre of the mid-oceanic ridge.

The abyssal plain is at the bottom of the continental slope. It is covered with a layer of mud, sand and animal remains.

The continental shelf is on the far edge of a continent, where the land slopes away to meet the ocean floor. It lies about 200 metres below sea-level.

The trench is a long, narrow valley which cuts through the ocean floor.

The mid-oceanic ridge is found on the abyssal plain. It is a chain of mountains which runs along the centre.

The continental slope is a steep cliff which drops about 3,658 metres to the ocean floor.

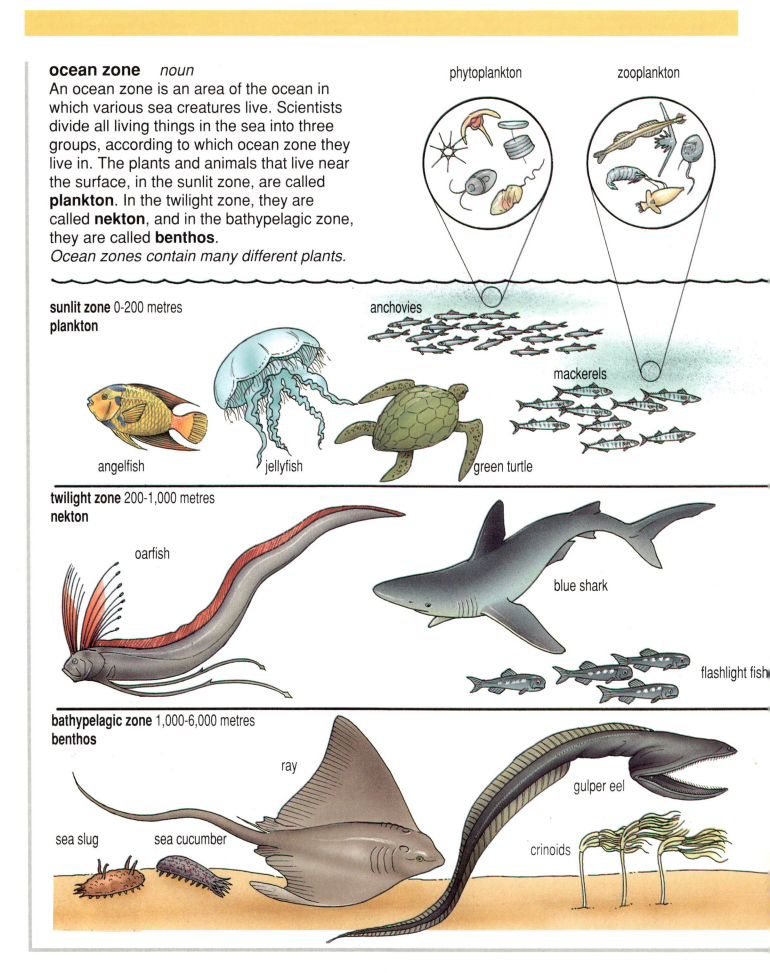

ocean zone *noun*

An ocean zone is an area of the ocean in which various sea creatures live. Scientists divide all living things in the sea into three groups, according to which ocean zone they live in. The plants and animals that live near the surface, in the sunlit zone, are called **plankton**. In the twilight zone, they are called **nekton**, and in the bathypelagic zone, they are called **benthos**.

Ocean zones contain many different plants.

phytoplankton

zooplankton

sunlit zone 0-200 metres
plankton

anchovies

mackerels

angelfish

jellyfish

green turtle

twilight zone 200-1,000 metres
nekton

oarfish

blue shark

flashlight fish

bathypelagic zone 1,000-6,000 metres
benthos

ray

gulper eel

sea slug

sea cucumber

crinoids

gulls

frigatebird

booby

ing fish

basking shark

dolphin

sei whale

tuna

octopus

sperm whale

giant squid

deep-sea eel

angler fish

devilfish

tripod fish

brittlestars

sea lilies

oil platform *noun*

An oil platform is a structure that supports underwater drilling equipment. In shallow water, drilling and pumping platforms are fixed to the sea-bed by huge steel legs. In deeper water, floating platforms are used. These are fixed to the sea-bed by cables. They usually have a landing pad for helicopters, and a hostel for the workers to stay in. Platforms used for collecting and storing oil have underwater storage tanks. *The helicopter arrived to take the oilmen to the oil platform in the North Sea.*

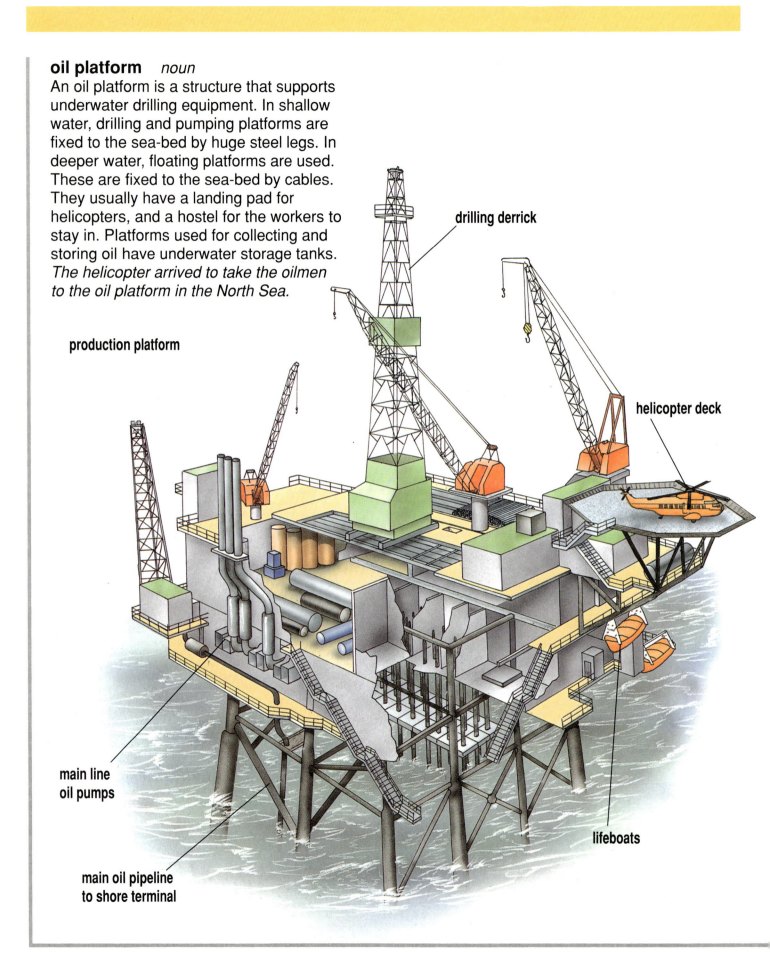

production platform

drilling derrick

helicopter deck

main line oil pumps

main oil pipeline to shore terminal

lifeboats

The template rig is used in shallow to medium water depth.

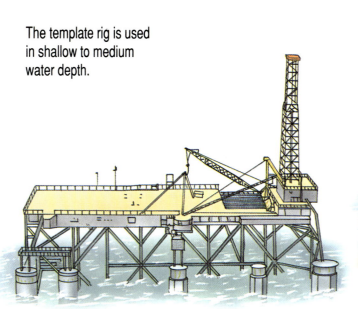

The semisubmersible rig is used in deep water operation.

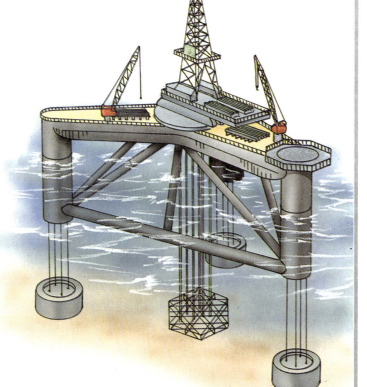

The jack-up rig is used in water of a medium depth.

The tension-leg rig is used in experimental development.

oxygen *noun*
Oxygen is an **element** that makes up one-fifth of the Earth's atmosphere. It is a colourless gas which is also found in sea water. Animals need oxygen in order to burn up their food to release energy. Plants produce oxygen during **photosynthesis**. Animals breathe in oxygen and use it up.
The seaweeds in the rock pool gave off bubbles of oxygen.

Oyashio Current *noun*
The Oyashio Current is a **surface current** in the North **Pacific Ocean**. It carries cold Arctic water south-west along the coasts of the Kamchatka Peninsula and the Kuril Islands of Russia. Then it sinks below the warm Kuroshio Current near Japan and continues south as a deeper current.
There is thick fog where the Oyashio Current meets the warm Kuroshio Current.

oyster ► **bivalve**

oystercatcher *noun*
An oystercatcher is a large black-and-white **shorebird**, which feeds on **bivalves** such as **mussels** and **cockles**, and on **worms** that live on the **shore**. Oystercatchers have long, strong, orange beaks, which they use to prize open the bivalve shells or hammer them to pieces.
As the students approached the rocks, they startled two oystercatchers who flew up with loud alarm calls.

ozone layer *noun*
The ozone layer is the name for part of the Earth's upper atmosphere. It contains large quantities of the gas ozone, which is a special kind of oxygen. The ozone layer protects the Earth from the harmful radiation of the Sun. Certain gases called chlorofluorocarbons, or CFCs, can damage the ozone layer.
If the ozone layer gets thinner, the effects of sunburn will become far more damaging.

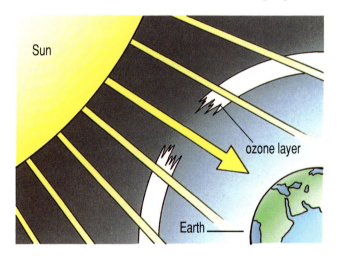

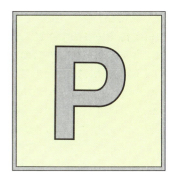

Pacific Ocean   *noun*
The Pacific Ocean lies between Asia, Australia, Antarctica, North America and South America. It is the world's largest and deepest ocean. There are many **islands** and **coral reefs** in the tropical areas of the Pacific Ocean.
One third of the Earth's surface is covered by the Pacific Ocean.

pack ice   *noun*
Pack ice is the floating mass of ice that forms when the sea freezes over in polar regions.
The polar bears hunted for seals on the pack ice.

palolo worm   *noun*
A palolo worm is a relative of the **ragworm**. It lives on the sea bed in the **Atlantic** and **Pacific Oceans**. When palolo worms are ready to reproduce, the back sections of their body which contain the eggs and sperm, break off and swim to the **ocean** surface. This happens at exactly the same time each year, eight to nine days after full moon in November.
The people of Fiji and Samoa collect and eat the eggs of palolo worms.

Panama Canal   *noun*
The Panama Canal is a man-made waterway which cuts through the **Isthmus of Panama**. It is 81.63 kilometres long. The Panama canal links the Atlantic and **Pacific Oceans**.
More than 10,000 ships a year use the Panama Canal.

parasite   *noun*
A parasite is an **organism** that lives in or on another organism, and feeds on it.
Many lampreys are parasites of other fish.

parrotfish   *noun*
Parrotfish are brightly coloured **coral reef fish** which have a horny beak instead of teeth. They feed on **corals**.
There are almost 100 species of parrotfish.

passenger ships   ► page 104

pearl   *noun*
A pearl is a small ball of **mother-of-pearl** which has been formed inside the shell of an **oyster** or other kind of **bivalve**. An oyster deposits a **mucus** which hardens into mother-of-pearl around any strange object, such as a small piece of grit, that gets trapped inside its shell. A cultured pearl is made by placing a tiny bead inside an oyster to make it produce a pearl.
A giant clam produced the largest pearl in the world, valued at over $33,000,000.

pearl

pearlfish   *noun*
A pearlfish is a small, tropical **fish** with a large head and a thin tapering tail. Some pearlfish live in **symbiosis** with certain **sea cucumbers**, **bivalves** and **starfish**, or host animals. The pearlfish enters its host's body tail-first, through its anus, and stays there by day. It leaves its living home to feed at night.
Most pearlfish do their hosts no harm, but a few actually feed on their tissues.

passenger ship   *noun*

A passenger ship is a vessel designed to carry passengers across short distances of water. It has been replaced by the aeroplane for carrying people on long trips.
Passenger ships include hydrofoils and hovercraft.

Roll-on, roll-off ferries carry cars and people.

Passenger liners are mainly used by holiday-makers, for cruises across the ocean.

Hovercrafts provide fast trips over short distances. They carry people and vehicles across areas of water such as the English Channel.

Hydrofoils can provide high-speed transportation over short distances of water.

pelican *noun*
A pelican is a large bird. It has short legs, webbed feet, a plump abdomen, short tail and long wings. A pelican has a large, pointed beak attached to a stretchy throat sac. The pelican uses its beak to scoop fish into its throat sac before swallowing them.
A pelican was diving in the bay.

penguin ► page 106

peninsula *noun*
A peninsula is a strip of land which is almost completely surrounded by water.
The Peloponnese Peninsula is joined to the Greek mainland by the Isthmus of Corinth.

Persian Gulf ► **Arabian Gulf**

Peru Current *noun*
The Peru Current, or Humboldt Current, is a branch of the cold **West Wind Drift**, flowing north up the west coast of South America.
Many food fish can be found in the Peru Current.

petrel *noun*
A petrel is a bird belonging to a group of seabirds called the tubenoses. Its nostrils are extended as horny tubes along the top of its beak. There are several different kinds of petrel. Storm petrels snatch fish from the surface of the sea. Diving petrels swim after their prey under water.
When threatened, petrels squirt a stinking oil at the intruder.

petroleum *noun*
Petroleum is a thick, dark liquid that gives off plenty of heat when it is burned. It is a fossil fuel because it was formed over millions of years from the remains of plants. Petroleum and gas are often trapped together in the same underground rocks. They are reached by drilling a hole through the rock, then inserting pipes to carry them to the surface. Once the hole is full of petroleum, it is called a petroleum well. Petroleum is sometimes called oil.
The oilmen were drilling holes in the sea bed in search of petroleum.

Philippine Sea *noun*
The Philippine Sea is part of the **Pacific Ocean**. It lies north and east of the Philippine Islands. It is surrounded by the Philippine and Caroline Islands, the Marianas, the Bonin and Volcano islands, Japan, the Ryukyu Islands and Taiwan.
During the Second World War, a great naval battle was fought in the Philippine Sea.

phosphorescent sea *noun*
A phosphorescent sea is a sea with a luminous surface that gives off flashes of light. The light is produced by millions of tiny **dinoflagellates** living in the **plankton**. When disturbed by water movements, they produce flashes of light.
Lights appeared to dance on a phosphorescent sea around a boat that was moving.

photosynthesis *noun*
Photosynthesis is the name of the process used by green plants and **algae** to produce their own food using the energy of sunlight. They take in simple substances such as water, **carbon dioxide** and minerals and join them together to make new, living material. Oxygen is given off as a waste product.
The plankton algae trap energy during photosynthesis and pass it on to other creatures in the food chain.
photosynthesize *verb*

penguin *noun*

A penguin is a large **seabird**. There are 16 **species** of penguin, all found in the southern hemisphere, mainly in Antarctica and around the southern coasts of South America and Australia. They feed mainly on fish, **squid**, small **crabs** and **krill**. Penguins cannot fly. Their wings look like paddles and they use them for underwater swimming.

Penguins have short, thick feathers which are waterproof and protect them from the cold.

The jackass penguin lives off the coast of South Africa. It is so called because it brays like a jackass or a donkey.

The emperor penguin is the largest and heaviest of all penguins. It can be as tall as 2 metres.

The Galapagos penguin can be as small as 50 centimetres high. It lives only on the Galapagos Islands.

The fairy penguin, or little penguin, is the smallest of the penguin species. It is about 40 centimetres tall and weighs only 1-1.5 kilograms. The fairy penguin is found in southern Australia and New Zealand.

The rockhopper penguin lives in the Falkland Islands. It is about 64 centimetres tall.

phytoplankton   *noun*
Phytoplankton is the part of **plankton** that traps the Sun's energy by **photosynthesis**. Most of phytoplankton consists of **algae**. Phytoplankton can only grow using the process of photosynthesis. Phytoplankton serves as food for **zooplankton** and other larger marine animals.
The larvae of many marine invertebrates feed on phytoplankton.

pier   *noun*
A pier is a wooden or steel platform that runs from a shore-line out into the sea. Piers are supported by posts, set into the sea-bed, deep enough to sit on a firm base.
Piers are built for commercial purposes so that large ships that need a deep berth can tie up alongside to be loaded and unloaded.

pilchard ► **herring**

pillow lava   *noun*
Pillow lava is the name given to volcanic magma once it has cooled and hardened under water to form rocks shaped like pillows.
The stream of hot lava hissed as it cooled and then it slowly crumpled to form pillow lava.

pilot ship   *noun*
A pilot ship is a small ship that guides larger ships into a **harbour** or port. It leads them safely between any rocks, reefs or **sand-banks**.
The huge tanker waited for a pilot ship to guide it up the estuary.

pilotfish   *noun*
A pilotfish is a pale blue fish with five or six dark stripes. It is found near the surface of the water in most tropical seas. Pilotfish swim alongside ships and large fish, like a **pilot ship**. They feed on scraps of food, and on the **parasites** of larger fish. Pilotfish are about 30 centimetres long.
Two pilotfish were swimming with the shark, keeping well out of reach of its jaws.

Pinnipedia ► page 108

pipefish   *noun*
A pipefish is a long, narrow **fish** with a stiff body covered in bony plates. It has a long, tubular snout, and is well adapted for gliding into narrow cracks and crevices in search of food. Pipefish are related to **seahorses**. Male pipefish have a pouch in which eggs are incubated. One of the largest pipefish is the ocean pipefish, which grows to over 60 centimetres long.
Pipefish can change their colour to match their surroundings.

pistol shrimp   *noun*
A pistol shrimp is a small **shrimp**, which has one extra large pincer. On one side of the pincer is a knob which fits into a socket on the other side. As the pincer is snapped shut, it makes a loud noise like a small pistol shot.
When pistol shrimps want to threaten or show off to each other, they snap their pincers.

plaice ► **flatfish**

plankton   *noun*
Plankton describes the floating, microscopic **organisms** that live in oceans, seas, lakes and ponds. Plankton includes tiny **crustaceans**, the larvae of many marine creatures, fish fry, and **algae** such as diatoms and **dinoflagellates**.
Shoals of fish follow floating plankton to feed off it.

pinniped *noun*

A pinniped is a mammal belonging to the pinnipedia group. Pinniped means 'fin feet'. There are three main groups of pinniped. These are eared seals, which include fur seals and sealions, earless seals which include harbour seals and elephant seals, and lastly walruses. Pinnipeds spend most of their time in water, but are also able to come on land using their flippers. They have oily fur, and a thick layer of **blubber** under their skin. These help keep them warm.
All pinnipeds are meat-eaters, or carnivores.

earless seals

ringed seal

eared seals

Californian sealion

harbour seal

northern fur seal

elephant seal

Arctic walrus

The walrus has thick, tough hide which is thinly covered in dark brown hairs. It has a thicker layer of blubber than seals, because it has no fur.

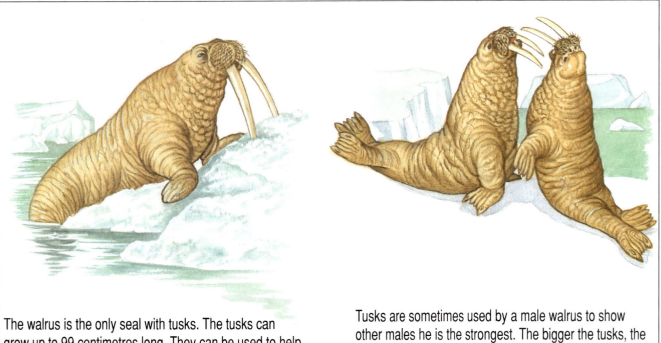

The walrus is the only seal with tusks. The tusks can grow up to 99 centimetres long. They can be used to help the adult walrus haul itself out of the water onto the ice.

Tusks are sometimes used by a male walrus to show other males he is the strongest. The bigger the tusks, the more superior the walrus.

plate tectonics *noun*
Plate tectonics is a scientific theory which states that the Earth's crust is formed by a number of large **tectonic plates**. These plates are moved slowly around the planet by currents in the molten rocks of the mantle. When two plates collide, one plate may be forced underneath the other. Or the plates may squeeze up layers of sediments to form new mountain ranges. Where plates move apart, **sea-floor spreading** takes place.
The study of plate tectonics shows how the continents have moved over thousands of years.

plesiosaur *noun*
A plesiosaur was a long-necked, prehistoric marine reptile that lived 190 to 65 million years ago. It had a pointed tail and paddle-like limbs. Plesiosaurs were agile hunters that fed on **fish** and **squid**.
Plesiosaurs probably used their long necks to raise their head out of the water to scan the waves for signs of prey.

pliosaur *noun*
A pliosaur was a prehistoric marine reptile, rather like a **plesiosaur**, but with a shorter neck, large head and powerful jaws. Pliosaurs could swim very fast, and hunted **sharks**, large **squid** and even **ichthyosaurs**. Pliosaurs ranged from about 4.5 metres to over 12 metres in length.
The heads of some pliosaurs were almost three metres long.

poison *noun*
Poison is a substance that can kill living things or make them very ill. Some fish and **shellfish** contain poisons which they use for defence. Such creatures are usually brightly coloured as a warning. Certain **dinoflagellates** release poisons into the water. They then poison fish and shellfish and the humans who eat them.
Jellyfish release poison in their tentacles.
poison *verb*

polar *adjective*
Polar describes something which comes from, or which is found close to, the North Pole or the South Pole.
The north wind sweeps cold, polar air across Siberia.

polar bear *noun*
The polar bear is a large white bear that lives in Arctic regions. It feeds mainly on **seals** living on the ice around the shores of the **Arctic Ocean**. The polar bear has waterproof fur. It is a very strong swimmer.
Wild polar bears can live up to 33 years old.

polder *noun*
A polder is an area of land that was once covered by the sea, but has now been reclaimed, or turned into dry land. Polders are made by building walls around the area, then either draining it or pumping the water out of it.
Many Dutch farms can be found on polders.

poles *noun*
The poles of the Earth are the points at its extreme north and south, at **latitude** 90 degrees. The North Pole and the South Pole are directly opposite each other. The Earth spins around an imaginary line known as its axis which is drawn between the two poles. Both the poles are covered in ice. The North Pole is over the frozen **Arctic Ocean**, while the South Pole is over Antarctica.
At the pole, they found flags left by earlier explorers.

pollution ▶ page 112

polyp ▶ **cnidaria**

polynya *noun*
A polynya is an area of open water surrounded by ice.
Many seals and walruses spend the Arctic winter in polynyas.

pompano *noun*
Pompano is the name given to several species of salt water fish. Pompanos are valuable food fish. They travel in **schools** close to the shore. The body of a pompano is very deep and flattened and its mouth is very small.
Pompanos are usually caught in nets.

porcupinefish *noun*
A porcupinefish is a spiny fish that puffs itself up like a balloon when danger threatens. It does this by swallowing air or water. This makes the porcupine fish look larger than it really is. It also makes it very difficult to swallow. **Pufferfish** act in the same way.
The spines of porcupinefish and pufferfish stick out when they puff up their bodies.

porpoise *noun*
A porpoise is a fish-shaped mammal that looks like a small whale. It has a rounded snout, and a small number of flat teeth.
Porpoises feed on small fish and squid.

port *noun*
1. A port is a place on the coast which provides shelter for ships. Ports are places where ships load and unload goods or fish, and where ferries pick up and set down passengers.
The ships that were carrying oil docked at the port.
2. Port is the side of a ship which is on the left when a person faces the front of the ship. The opposite of port is **starboard**.
In order to sail through the narrow channel, the captain had to steer the ship to port.

Portuguese Man-of-War *noun*
The Portuguese Man-of-War is a sea creature that looks like a large **jellyfish**. It lives at the surface of the ocean. It has a large, gas-filled float which catches the wind to move it along. The Portuguese Man-of-War is not a true jellyfish. It is a colony of **polyps**, each with a different job. One polyp forms the float. Other polyps provide the **tentacles**, digest food or produce new men-of-war.
The tentacles of the Portuguese Man-of-War are up to 20 metres long and highly poisonous.

Poseidon *noun*
Poseidon was the Greek god of the **sea** and earthquakes. He was also identified with the Roman god **Neptune**. Poseidon is often shown with a **dolphin** and a three-pronged fork with a long handle, or trident.
Poseidon was said to be able to change his form in order to court young women.

pollution *noun*
Pollution describes the adding of unwanted substances to the environment. This upsets ecosystems and poisons air, water and soil. Gases and other chemicals also dissolve in lakes, river and oceans. Particles of waste cloud the water and choke coral reefs.
Pollution has made many beaches unsafe to bathe in.

Sewage uses up oxygen in the water and smothers marine life.

Oil spilled from tankers and the shore covers birds in oil and stops them from flying. When they try to clean their wings, the oil poisons them.

Corals bleach, or turn white and may die if they come into contact with pollutants or if the temperature of the sea increases.

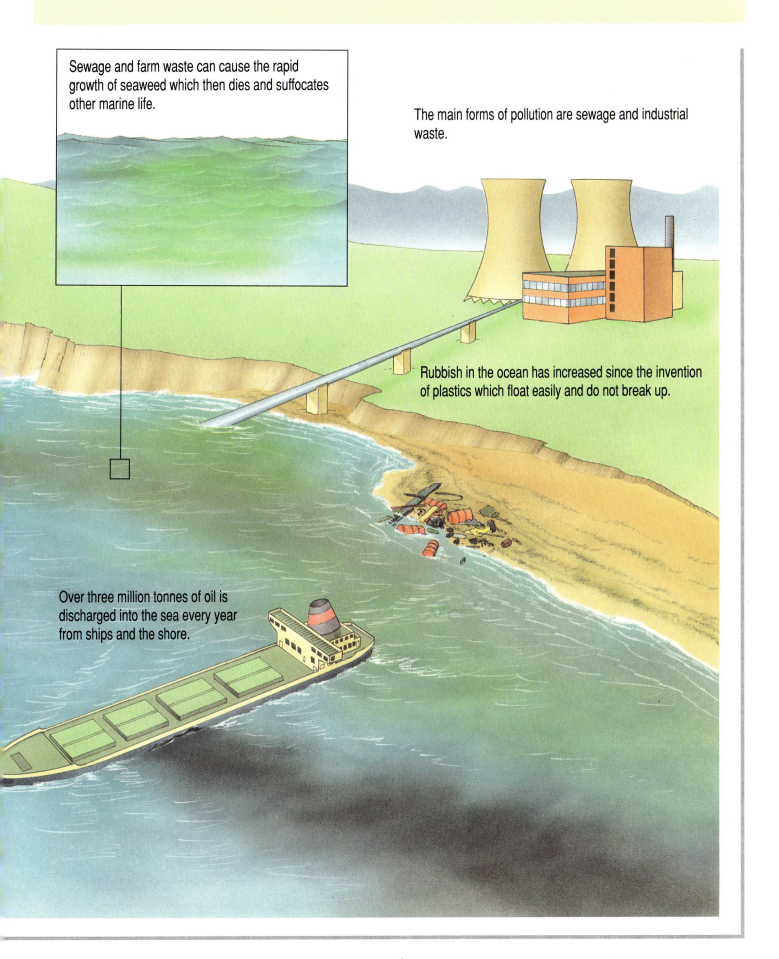

Sewage and farm waste can cause the rapid growth of seaweed which then dies and suffocates other marine life.

The main forms of pollution are sewage and industrial waste.

Rubbish in the ocean has increased since the invention of plastics which float easily and do not break up.

Over three million tonnes of oil is discharged into the sea every year from ships and the shore.

113

prawn *noun*
A prawn is a **crustacean**. It has a humped back, long feelers, small claws, paddle-like legs and eyes on stalks. Prawns are usually about 12 centimetres long. Some tropical species can measure 30 centimetres long. Prawns are usually grey-coloured, with flecks of brown. They turn pink when they are cooked. Prawns are found in **rock pools** and in fresh water in warm temperate regions throughout the world.
Prawns are widely eaten and form an important part of the fishing industry in many countries.

precipitation *noun*
Precipitation is the name given to water falling from the sky as rain, snow, sleet or hail.
In Antarctica, most of the precipitation falls as snow.

predator *noun*
A predator is an animal that hunts other animals, or **prey**, for food.
Sharks are fierce predators.

pressure *noun*
Pressure describes the weight of something pressing down on something else. On the deep sea floor, the pressure of the water above is enormous. The weight of the atmosphere also creates a pressure, which is called atmospheric pressure.
Humans cannot survive the water pressure on the deep ocean floor.

prevailing wind *noun*
A prevailing wind is the wind that blows in a particular area most often.
At different seasons of the year, the direction of the prevailing wind may change.

prey *noun*
Prey describes an animal that is hunted by another animal called a **predator**.
The octopus uses its tentacles to seize its prey.

proboscis worm ► **ribbon worm**

protozoan *noun*
A protozoan is a tiny animal which is made up of only one cell. This cell is very complex. **Ciliates**, **foraminiferans** and **radiolarians** are protozoans.
Many protozoans feed on microscopic algae in the plankton.

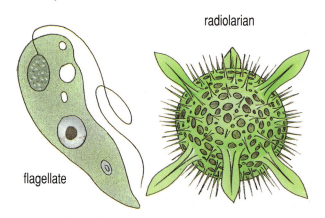

radiolarian

flagellate

prow *noun*
The prow is the front part of a ship. It often curves out over the water.
Many old sailing ships had beautifully carved prows.

pufferfish *noun*
A pufferfish is a small fish. There are about 120 species of pufferfish. They live mainly in warm tropical seas and on **coral reefs**. Pufferfish have a hard, beak-like mouth. When they are threatened, they can blow up their body with water until it is much bigger than its normal size.
Many species of pufferfish have a poisonous body.

puffin ► **auk**

purse seine net *noun*
A purse seine net is a curving net with weights at the bottom and floats at the top. Once the fish are inside the net, the fishermen pull a steel cable to close it like a bag and pull it into the ship.
Fish that swim in shoals near the ocean surface are caught using purse seine nets.

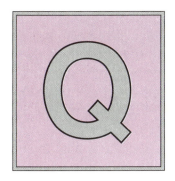

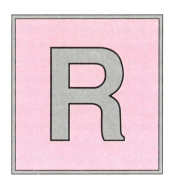

quahog *noun*
A quahog is an edible clam with a colourful shell. The ocean quahog is the longest-living **mollusc** in the world. In fact, the oldest known quahog is 220 years old.
The American Indians used to make purple beads from the shells of quahogs.

quicksand *noun*
Quicksand is very wet **sand** that becomes even more liquid when it is pressed. It is easy to sink into quicksand. Quicksand is found in some **estuaries** and on some **beaches**.
The boy lost his boots when they pulled him out of the quicksand.

rabbitfish *noun*
A rabbitfish has a rounded head and a rabbit-like mouth. Its **fins** have spiny, poison-tipped rays, which produce painful stab wounds which are slow to heal. Like rabbits, rabbitfish are grazers. They feed on plants and algae in shallow water in the tropical **Pacific Ocean**. There are about 24 species of rabbitfish.
In parts of the Pacific, rabbitfish are considered to be a delicacy to eat.

radioactive waste *noun*
Radioactive waste is material discharged into the **oceans**, which gives off harmful radiation. Many nuclear power stations are found along coasts, where they discharge their cooling water directly into the **sea**. The radiation damages marine life, and can be harmful to bathers.
Much of the radioactive waste in the Irish Sea comes from the nuclear waste reprocessing plant at Sellafield, in the United Kingdom.

radiolarian *noun*
A radiolarian is a **microscopic** animal that lives in **plankton**. Each radiolarian is a spherical **cell** surrounded by a glassy shell covered in long spines. Long strands of living material reach out from the tips of the spines, to trap floating food **particles**. Radiolarians have a tiny skeleton which is made up of a substance called silica. After a radiolarian dies, the skeleton sinks to the deep ocean floor.
When radiolarian skeletons build up, thick layers of deposits form, called oozes.

raft   *noun*
A raft is a floating platform. It is often made from pieces of wood or bundles of reeds bound together. Modern inflatable rafts are made of rubber bags pumped full of air.
The helicopter dropped an inflatable raft to rescue the drowning sailors.

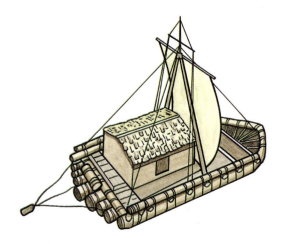

ragworm   *noun*
A ragworm, or **sandworm**, is a segmented **worm** that lives on sandy and muddy **shores**. It hunts for worms and other small animals while the tide is in, using bristly paddles on its segments for swimming.
She found a ragworm on the seashore.

raised beach   *noun*
A raised beach is a **beach** that used to be covered by the sea but which now lies above the **high water mark**. Since it was first formed, either the land has risen or the **sea-level** has fallen.
The children went looking for shells on the raised beach.

Ramsar Convention   *noun*
The Ramsar Convention is an agreement between some countries to protect important wetland habitats, such as marshes and swamps, and their wildlife. Ramsar is also known as the Convention on Wetlands of International Importance.
The lagoons and mudflats of the Coorong National Park in Australia are protected under the Ramsar Convention.

rat-tail fish   *noun*
A rat-tail fish, or grenadier, is a **deep sea fish** with a long, tapering, scaly tail. It has large eyes for seeing in the dark waters. Many **species** of rat-tail fish have a **barbel** on the chin, which they use to feel for food on the sea-bed. The mouth is on the underside of the snout and is used for feeding.
Rat-tail fish are some of the commonest fish in the deep sea.

ratfish ► chimaera

ray   *noun*
A ray is a **cartilaginous fish,** with a wide, flattened body and a long tail. The giant manta ray is over six metres across. Most rays feed on **invertebrates** on the sea-bed, so their mouth is on the lower part of their body. Electric rays stun their prey with electric shocks. Stingrays have sharp, poisonous spines on their tails, for defence.
Rays swim by flapping their broad pectoral fins up and down.

razor shell   *noun*
A razor shell is a long, narrow **bivalve**. The shells are very smooth and have straight sides. Razor shells use their muscular foot to burrow in the **sand**. They are found almost everywhere, except on dry land.
The men were digging on the beach to harvest the tasty razor shells.

razorbill ► auk

Red Sea   *noun*
The Red Sea is a narrow part of the **Indian Ocean**, found between north-east Africa and the Arabian Peninsula. It is linked to the **Mediterranean Sea** by the **Suez Canal**.
Many corals grow in the Red Sea.

red seaweed   *noun*
Red seaweed is a delicate seaweed which is red, purple or greenish-purple in colour. Most red seaweeds live in cooler, deeper water than other kinds of seaweed. Only a few live in **rock pools**. Many red seaweeds have branching, feathery fronds. Some form flat or curly sheets. Laver and Irish moss are collected and eaten. Coralline red seaweeds are covered in a hard crust of **calcium carbonate**. They help to build up reefs.
Some red seaweeds can survive at depths of over 60 metres.

red tide   *noun*
A red tide describes the reddish colour in the **sea** when large numbers of tiny **dinoflagellates** are present in the water. The dinoflagellates may poison **fish** and **shellfish**. Red tides are common in warm seas.
Hundreds of dead fish were washed up on the beach after the red tide.

redfish ► **scorpionfish**

reef   *noun*
A reef is a ridge of rock, **coral**, **shingle** or **sand** lying just above, or just below, the surface of the **sea**.
The ship was found wrecked on the reef.

remora   *noun*
A remora is a slim **fish** with a sucker on the top of its head with which it attaches itself to the abdomen of **sharks, whales** and **sea turtles**. Remoras live in warm or tropical seas. They measure about 15 to 110 centimetres long. Some remoras attach themselves to ships' hulls or other floating objects.
The sharksucker is a large remora, measuring up to one metre long.

remote sensing   *noun*
Remote sensing is a method of studying the Earth's natural features from a great height. Special cameras and other instruments are carried by satellite and are aimed at the feature to be studied. Some of the cameras may be infra-red, so that a detailed picture of the area may be built up.
They used remote sensing equipment to map the ocean floor.

reptile   *noun*
A reptile is a **vertebrate** whose body is covered in horny scales. There are about 6,000 species of reptile. Reptiles are 'cold-blooded', which means that their body temperature varies according to the temperature of their surroundings. They must avoid very high or very low temperatures to stay alive. Most reptiles lay eggs, but a few give birth to well-developed young.
Turtles, lizards, alligators and snakes are examples of reptiles.

research ship   *noun*
A research ship is a ship designed to carry equipment and people to study life and conditions under the ocean. It may be used to monitor sea currents, plant or animal life, or to map the ocean floor. Research ships usually have a laboratory on board where samples that have been collected from the ocean can be studied.
The marine biologists were studying on board the research ship.

ria *noun*
A ria is a valley which has been drowned because the **sea-level** has risen or the land has sunk. It produces a funnel-shaped **estuary** with deep water around its mouth.
The sheltered rias of the Cornish coast, England, are popular places for sailing.

ribbon worms *noun*
A ribbon worm, or **proboscis worm**, is a very thin worm, which captures its **prey** with a proboscis, or a contracting tube with a pointed end. Ribbon worms live in **mud** or under rocks by day, and come out at night in search of other worms to eat. The ribbon worm shoots out the front part of its proboscis and curls it around its prey.
It covers the prey in sticky **mucus**, then injects poisons into it.
Some ribbon worms can stretch themselves out to lengths of 27 metres.

Ring of Fire *noun*
The Ring of Fire describes the area surrounding the **Pacific Ocean**, where there are many volcanoes. The **tectonic plates** which carry the expanding Pacific Ocean floor, are being forced under the plates carrying the surrounding continents. The pressure that results often causes cracks in the Earth's crust. **Magma** forces a path through, erupting on the ocean bed. This violent action produces many earthquakes and volcanoes.
The volcanoes of the Philippines form part of the Ring of Fire.

rip current *noun*
A rip current is a strong current that flows outwards from the **shore**. When waves approach land at an angle, a current flows sideways along the shore. The rip current carries this water out to sea. Rip currents often happen where the water can escape through a gap in a **sandbank** or **reef**.
The dinghy was caught in the rip current.

Roaring Forties ► Forties

rock pool *noun*
A rock pool is a hollow area found in rock which has filled up with water left behind by the tide. When the tide is out, rock pools may become hot and salty as the water **evaporates**. After heavy rains they may contain almost fresh water.
Many small animals such as starfish and crabs and different kinds of seaweed live in rock pools.

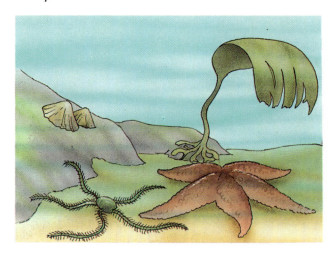

rockling ► cod

rocky shore *noun*
A rocky shore has very little **sand** or **mud**. There are **rock pools** in hollows in the rocks. Below **high water mark**, the rocks are covered in **seaweeds**. Animals of rocky shores, such as **limpets** and **winkles**, cling hard to the rocks. This helps to stop them being washed away by the **waves** when the tide is in. Animals which do not have hard shells shelter under seaweeds or hide in cracks in the rocks.
Starfish and crabs live on rocky shores.

RoRo ► ferry

Ross Ice Shelf *noun*
The Ross Ice Shelf is the largest area of floating ice in the world. It is about the size of France. In places, the ice is more than 700 metres thick.
The Ross Ice Shelf lies at the head of the Ross Sea on the shores of Antarctica.

Ross Sea   *noun*
The Ross Sea is part of the south **Pacific Ocean**. It is between the Victoria and Edward VII Peninsulas on the coast of Antarctica. The **Ross Ice Shelf** forms the southern border of the Ross Sea.
Many volcanoes, old and new, rise from the shores of the Ross Sea.

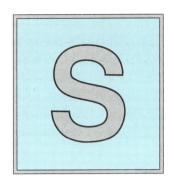

sailfish   *noun*
A sailfish is a large, open ocean fish, with an extra-long upper jaw shaped like a spear. Sailfish use their bills to slash and stun **prey**, which they attack at speed. They have a very large dorsal **fin**, which they can raise like a sail. This fin can be folded into a groove on the sailfish's back to improve streamlining while swimming.
Sailfish are the fastest fish in the sea, reaching a speed of 109 kilometres per hour.

sailing ship   *noun*
A sailing ship is a ship which uses sails. In the past, nearly all ships used sails. By the mid 1800s most sailing ships had been replaced by steam ships, which could travel faster.
Sailing ships were often used to carry passengers across the ocean.

saline   *adjective*
Saline describes a liquid which contains **salt**. **Sea water** is a liquid which is naturally saline, because it contains huge quantities of common salt.
Objects float more easily in saline water than in fresh water.

119

salinity *noun*
Salinity describes how much **salt** there is in a sample of water. Salinity is a measure of the amount of mineral salts dissolved in the water. It is usually measured in parts per thousand. The open ocean usually contains about 35 parts of salt per thousand parts of water.
The water in the oceans has a higher salinity than the water in rivers.

salinometer *noun*
A salinometer is an instrument that measures the amount of **salt**, or sodium chloride, in a sample of sea water.
The oceanographer tested the sample of sea water with a salinometer.

salmon *noun*
A salmon is a long, narrow fish. It has a small, fleshy **fin** on its back, behind the main dorsal fin. Salmon live in cool waters. They spend most of their adult life at **sea**, but **migrate** to rivers to **spawn**. Young salmon return to the sea once they reach a certain size.
Small salmon feed on invertebrates and larger salmon feed on other fish.

salp *noun*
A salp is a transparent, barrel-shaped animal that floats in the **ocean**. Salps pump water through their body and filter out **microscopic** organisms. The water that is squirted out of the rear of the salp helps to propel it along. Some salps live alone, while others line up in long chains.
The long chain of salps were almost invisible against the sunlit water.

salt *noun*
Salt, or common salt, is the popular name for the chemical **sodium chloride**. Solid salt forms white, cube-shaped crystals. These are left behind when salt water evaporates. There are about 100 grams of salt in five litres of sea water. Salt that comes from evaporated sea water is sometimes called solar salt.
Common salt makes the sea salty.

salt gland *noun*
A salt gland is a structure which is used to get rid of unwanted salt. **Sea turtles** cry salt tears. These are produced by a salt gland which removes salt from the turtles' blood.
The sea lavender was covered in salt given out from its leaves by salt glands.

salt water *noun*
Salt water is water which contains more than 1,000 parts per million of dissolved **salt**, although it can also describe any water that tastes salty. Salt water is denser than fresh water. When salt water meets water which is less salty, it will sink below it. This causes a movement and is how many deep ocean currents start.
It is easier to float in salt water than in fresh water.
saltwater *adjective*

Salter's Duck *noun*
A Salter's Duck is a device used for trapping, or harnessing, the energy of **waves**. Each 'duck', is a hollow float that rides on the waves. As it bobs up and down, the rocking motion produces energy which drives an electric generator.
The engineers tested a long line of Salter's Ducks on Loch Lomond in Scotland.

saltern *noun*
A saltern, or saltworks, is a group of artificial **saltpans** used to extract **salt** from **seawater**.
As the seawater in the saltern evaporates, the salt is left behind as a white crust.

salt-marsh   *noun*
A salt-marsh is an area of flat, wet ground which is sometimes flooded by **salt water**. Salt-marshes are common along the banks of **estuaries**, **deltas** and low sea **coasts**. They are usually covered in tough grasses and special kinds of plant called **halophytes**.
The geese come to graze on the salt-marsh.

saltpan   *noun*
A saltpan is a shallow hollow in the ground. Sea water **evaporates** in a saltpan, leaving behind a layer of salt. Natural saltpans are found on salt-marshes. Artificial saltpans are used to take out, or extract, salt commercially. Groups of artificial saltpans are sometimes called saltworks or salterns.
There was a rim of white salt around the saltpan.

salvage   *noun*
Salvage refers to the rescue of a ship, its cargo or its passengers from shipwreck or damage at sea.
People who help with salvage are entitled by law to payment.

sand   *noun*
Sand is made up of tiny particles of rock, mineral or soil that measure between 0.02 and 2 millimetres across. Sand particles are larger than mud particles.
Over hundreds of years, the waves ground the rocks down into sand.

sand dollar ► sea urchin

sand dune   *noun*
A sand dune, or dune, is a mound of **sand** that has been piled up by the wind. Sand particles are so light that they can be shifted easily by an air current. However, when the air current slows down, it drops the sand it is carrying. This often happens when an air current meets a solid object such as a rock or plant. Sand dunes may form behind sandy **beaches** where the prevailing wind blows off the **sea**. Sand dunes are also found in desert regions. Here, the wind blown sand covers large areas of land.
Many birds nest in shelter of the sand dunes.

sand-bank   *noun*
A sand-bank is a mound of **sand** which is found on the sea-bed, or on the bed of an **estuary** or **river**. A sand-bank forms when a water current, which is carrying sand, slows down. The sand sinks down through the water and is left, or deposited, as a mound.
The ship ran aground on a sand-bank which was hidden below the water.

sandhopper   *noun*
A sandhopper, or beach flea, is a small **amphipod** that lives on a sandy or a muddy beach. It looks like a flattened, curled-up **shrimp**. Sandhoppers hide in the **sand** by day, and come out at night to scavenge for food among the **seaweeds**.
As they walked along the strandline, sandhoppers leapt in all directions.

sandstone *noun*
Sandstone is a **sedimentary rock** formed from **sand** which has been cemented together by pressure or by minerals. The rock **particles** in sandstone consist of grains of feldspar, quartz and other minerals. They are between 0.0625 and 2 millimetres across. Sandstone may form when the sand in a large **delta** or sand bar, accumulates to a great depth. Sandstone can also form in deserts and in river beds.
That beautiful old church is made of red sandstone.

sandworm ► ragworm

sandy shore *noun*
A sandy shore has a **beach** of **sand**, often with **mud** near the low water mark. Animals, such as **cockles** and **marine worms**, burrow in the sand when the **tide** is out, and come out to **filter-feed** when the beach is covered in sea water. Others, such as **sandhoppers**, feed at low tide. They only do this at night when predators such as seagulls cannot see them. Seagulls scavenge along the **strandline**, and waders feed near the low water mark.
The children collected seashells on the sandy shore.

sardine ► herring

Sargasso Sea *noun*
The Sargasso Sea is a part of the North **Atlantic Ocean** which lies north-east of the **Caribbean Sea**. It is surrounded by the waters of the **Gulf Stream**, the Canary Current and the **North Equatorial Current**. Huge clumps of a brown seaweed called sargassum weed float in its warm water. This gives the sea its name. The sargassum weed provides shelter for many small animals like **crabs**, **shrimps**, **sponges** and **anemones**.
European eels travel all the way from their home rivers to the Sargasso Sea in order to spawn.

sargassum *noun*
Sargassum is a **brown seaweed**. It has fronds which have saw-like edges and large air bladders. Some sargassum **species** grow on rocks on the shore, but others float at the surface of the **ocean**. Large clumps of sargassum float in the **Sargasso Sea**, which is named after the seaweed.
Crabs, sea slugs and shrimps all live among sargassum.

satellite ► remote sensing

saturation diving *noun*
Saturation diving is a kind of diving used when work needs to be done at great **depths** under the **ocean**. The diver is pressurized in a pressure chamber on board ship. The pressure is the same as the pressure he will work at. When the diver has finished his work, he visits a **decompression chamber**.
They knew saturation diving was the only way they could repair the oil platform.

sawfish *noun*
A sawfish is a large, shark-like **ray**. It has a flat snout, which is narrow and edged with saw-like teeth. The saw is used to dig out animals from the sea-bed or to slash at shoals of **fish**.
Sawfish live in warm shallow waters, especially in bays and estuaries.

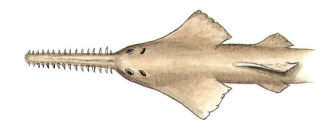

scale *noun*
A scale is a small, horny or bony plate. Scales cover the body of most fish and reptiles, and the legs of birds.
You can tell the age of some fish by counting the ridges on their scales.

scaleworm   *noun*
A scaleworm is a marine **worm**, which is covered in pairs of flat, overlapping scales. Its body is divided into many small sections. Scaleworms live as **predators** on the seashore and on the sea-bed.
Among the holdfasts of the kelps were several scaleworms.

scallop   *noun*
A scallop is a marine **bivalve**, which has a fan-shaped shell, with straight hinge lines. Scallops are often brightly coloured. They have lots of little eyes around the edges of their shell. A fringe of short **tentacles** hangs down between the valves when they are open. These detect changes in the chemicals in the water, which warn of the approach of enemies like **starfish**. Scallops swim by **jet propulsion** by clapping their shells open and shut.
They ate scallops for supper.

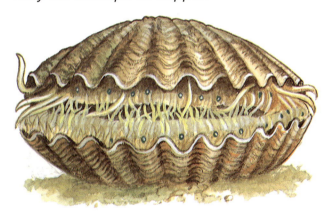

scavenger   *noun*
A scavenger is an animal that feeds mainly on the bodies of dead animals. **Crabs**, **sandhoppers** and hagfish are scavengers.
The dead fish had been almost completely eaten by scavengers.

school   *noun*
A school, or shoal, is a large number of fish or shrimps swimming together. School is also used to describe groups of whales, dolphins and porpoises.
A large school of porpoises was playing behind the boat.

scorpionfish   *noun*
Scorpionfish are a group of poisonous, spiny fish. The dorsal fins of scorpionfish have hollow spines which inject poison into anything they stab. Scorpionfish have good camouflage. Many can change colour to match their background and are able to lie in wait for their prey, hidden on the sea-bed. Lionfish, redfish and stonefish all belong to the scorpionfish family.
Divers must be careful not to step on scorpion fish, in case they get stung.

Scotia Sea   *noun*
The Scotia Sea is part of the south **Atlantic Ocean**. It lies south-east of the Falkland Islands. The Scotia Sea is bordered by the South Sandwich Islands, South Georgia and the South Orkney Islands.
The Scotia Sea was once an important centre for the whaling industry.

scuba diver   *noun*
Scuba diving is diving with the help of air tanks. The diver breathes from tanks of compressed air which are carried on the back. This allows the diver to stay under water for many minutes.
Scuba stands for self-contained under water breathing apparatus.

sea ► **ocean**

sea anemone ► **anemone**

sea bass ► **grouper**

sea butterfly   *noun*
A sea butterfly is a very small **gastropod** mollusc, which has a pair of wing-like flaps that it uses for swimming. Most sea butterflies are less than one centimetre long. Some have delicate, almost transparent shells, and trap food **particles** with bands of hair-like **cilia** on the 'wings'. Some sea butterflies have no shell.
Sea butterflies were swimming among the tiny animals of the plankton.

sea cow ► **sirenian**

sea cucumber   *noun*
A sea cucumber is a cucumber-shaped
echinoderm, with a soft or leathery skin.
It lives on the sea-bed, or buried in **mud** or
sand. At one end of a sea cucumber is a
mouth surrounded by branching **tentacles**.
These are used to collect **detritus**.
*Large numbers of sea cucumbers feed on
the detritus on the deep ocean floor.*

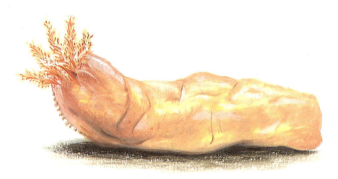

sea dragon ► **seahorses**

sea fan   *noun*
A sea fan is a brightly coloured, fan-shaped
coral, usually found in tropical seas.
*At night, the tiny, star-shaped polyps of the
sea fans came out to feed.*

sea floor ► **ocean floor**

sea floor spreading   *noun*
Sea floor spreading describes how new
ocean floor is formed. It is part of the
process of **plate tectonics**. Along the **mid-
oceanic ridges**, molten rock from deep in
the **mantle** is forced to the surface, pushing
the **tectonic plates** apart. Lava wells up
between them to form new ocean floor, so
the ocean floor spreads as a result.
The sea floor grows by sea floor spreading.

sea food   *noun*
Sea food is food obtained by humans from
the sea. It includes mainly fish and shellfish,
although some seaweeds are also eaten.
On the beach are stalls selling sea food.

sea goldfish   *noun*
A sea goldfish is a small, orange-red **fish**,
which forms large **shoals** on the **coral reefs**
of the Indian and **Pacific Oceans**. It looks
rather like goldfish, but it is really related to
groupers. Male sea goldfish have long
filaments trailing from their dorsal and
caudal **fins**.
Female sea goldfish have very large eyes.

sea gooseberry   *noun*
A sea gooseberry is a small transparent sea
animal, found in all the oceans of the world.
*Sea gooseberries move through the water
by beating bands of cilia.*

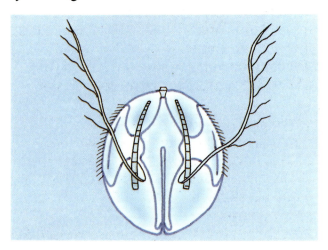

sea hare   *noun*
A sea hare is a slug-like **gastropod**. It has a
tiny shell which is hidden by two large, fleshy
flaps. These flaps are used like wings to
propel the sea hare through the water. Sea
hares feed on **algae**.
*When seen from the side, the tentacles of
a sea hare look like a hare's ears.*

sea lily   *noun*
A sea lily is an **echinoderm**. It looks a little
like a lily. A sea lily has a cup-shaped body
with five feathery arms surrounding its
mouth. It is fixed to the sea-bed by a long
stalk. The whole body is covered in chalky
plates. Sea lilies live mainly in deep water
where they feed on **detritus**.
*Fossils of sea lilies have been found in rocks
570 million years old.*

sea mat   *noun*
A sea mat, or moss animal, is a tiny animal that forms pale crusts over rocks, shells and **seaweeds**. Each animal is a cup-shaped polyp in a hard case. Often, dozens of these cases are joined together to form a colony, or mat. Each polyp has a horseshoe-shaped band of **tentacles** covered in hair-like **cilia**. Sea mats feed on **plankton**.
The white patches on the seaweed fronds were sea mats.

sea moth   *noun*
A sea moth is a small **fish**, with large pectoral **fins** that spread out from the sides of its body like wings. Its body is covered in bony plates, and it has a long tubular snout and a small mouth.
The body of a sea moth is so stiff that the fish has to rely on its 'wings' for swimming.

sea mouse   *noun*
A sea mouse is a large, broad **worm**. Its back is covered in long, grey bristles which shine in many different colours. The sea mouse lives near the low water mark on sandy shores.
The bristles on the back of a sea mouse look like the fur of a mouse.

Sea of Azov   *noun*
The Sea of Azov is an inland sea. It is found north-east of the **Black Sea**, between Ukraine to the west and Russia to the east. The Sea of Azov is linked to the Black Sea by the narrow Kerch Strait.
The Sea of Azov is very busy with passenger ships, cargo boats and fishing boats.

Sea of Cortez ► Gulf of California

Sea of Japan   *noun*
The Sea of Japan is part of the north-west **Pacific Ocean**. It lies between Japan and the mainland of Asia.
Natural gas and oil have been found beneath the Sea of Japan.

Sea of Marmara   *noun*
The Sea of Marmara is an inland sea, found in Turkey. It is linked to the **Black Sea** in the north-east, and to the **Aegean Sea** in the south-west.
The Islands of Marmara in the Sea of Marmara are famous for their marble.

Sea of Okhotsk   *noun*
The Sea of Okhotsk is part of the North **Pacific Ocean**. It lies between the Kuril Islands, the Kamchatka Peninsula and the east coast of Russia, and Japan's Hokkaido Island.
Ships crossing the Sea of Okhotsk have to cope with ice in winter and fog in summer.

sea otter   *noun*
A sea otter is a sleek, streamlined mammal which spends most of its life in the sea. Sea otters eat and sleep while floating on their back. They have thick brown fur that traps air and keeps them warm and dry. Sea otters can eat up to one fifth of their body weight every day.
Sea otters feed on crabs, clams, mussels, abalones, squid and sea urchins.

sea pen   *noun*
A sea pen is a soft, fleshy **cnidarian**, that looks like a quill pen. It is anchored to the sea-bed by a thick stalk. Each sea pen is really a colony of **polyps**.
On the deep sea floor, many sea pens glow in the dark.

sea potato   *noun*

A sea potato, or sand urchin, is a common **sea urchin** which lives on sandy **beaches**. It is about five centimetres across. The sea potato burrows in the **sand** with its tube feet, leaving a give-away hole at the surface.
As they removed the sand, they found a group of sea potatoes just below the surface.

sea robin   *noun*

A sea robin is a **fish** with a bony head and large, fan-shaped pectoral **fins**. The bottom few rays of the these fins form separate feelers, which are used for detecting **crustacean prey** in the **sediments**. They are strong enough to turn over rocks in search of food. These feelers, together with the stiff pelvic fins, help the fish to 'walk' over the sea-bed.
When the sea robin vibrated its swim bladder, using special muscles, it made a very loud noise.

sea slater   *noun*

A sea slater is a small **isopod** which looks like a woodlouse, or pillbug. Sea slaters are **scavengers** that live in the upper part of the shore. They hide by day, and come out to feed at night.
When she lifted the seaweed, several sea slaters scurried out.

sea slug ▶ page 127

sea snail ▶ **gastropod**

sea snake   *noun*

A sea snake is a very poisonous snake that lives in the **sea** and feeds on **fish**. Sea snakes have a flattened body and an oar-like tail to help them swim. They can close their nostrils to keep out the water. Some kinds of sea snake are also known as sea kraits.
The bright yellow and black stripes of the sea snake were a warning of its venomous bite.

sea spider   *noun*

A sea spider is a spider-like **arthropod** which lives on the sea-bed and in **rock pools**. Most sea spiders have four pairs of legs. They feed on soft-bodied **invertebrates**.
Sea spiders are good swimmers.

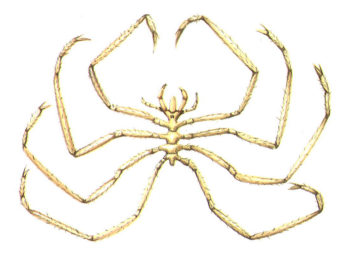

sea squirt   *noun*

A sea squirt is an **invertebrate** with a bag-like body. Each sea squirt has two tubes, called siphons. Water enters the body through one siphon and leaves through the other siphon. Internal filters sieve food particles from the water. Some sea squirts live alone. Others form colonies that spread over the surfaces of **corals** and rocks. Young sea squirts look like tadpoles.
Scientists think that sea squirts are related to ancestors of the vertebrates.

sea turtle ▶ page 128

sea urchin   *noun*

A sea urchin is a cushion-shaped **echinoderm**, covered in spines. The mouth is under the cushion and has five hard teeth. Urchins crawl over the sea-bed, or burrow in the **sand**. They feed on **algae** and soft-bodied animals such as **corals**. Their body is covered in five hard plates joined together, with small holes to let the tube feet through.
Sea urchins measure from 5 to 12 centimetres across.

sea slug   *noun*

A sea slug, or nudibranch, is a **gastropod** that has no shell and no gills. Sea slugs breathe through tissues shaped like small fleshy tufts on their back. Dorid slugs have these arranged in a circular pattern. Eolid slugs have fleshy spikes all over their back. Most sea slugs have poisonous flesh and many are brightly-coloured. They live in shallow water and feed on coral polyps, **sea anemones** and **sponges**.

Fish avoid sea slugs because of their nasty taste.

eolid sea slug

dorid sea slugs

sea turtle   *noun*

A sea turtle is a turtle that lives in the **sea**. It has legs like flippers which cannot support it on land. Sea turtles **migrate** to favourite beaches to breed. They mate offshore, and the females shuffle up the beach to lay their eggs. The warm sand helps the eggs to hatch, and the tiny turtles find their own way back to the sea. Sea turtles feed on **sea-grasses** or **algae**.

Many sea turtles are endangered and becoming rare, but they are still hunted for their meat and for their beautiful shells.

The leatherback turtle is the largest in the turtle family. It can be up to 1.8 metres long and can weigh up to 680 kilograms. The leatherback is able to feed on jellyfish because it is not affected by their sting. The leathery shell is grooved to aid fast swimming.

The loggerhead turtle has a very large head and powerful jaws. It eats mainly shellfish, sponges, fish and jellyfish.

The hawksbill turtle has a snout shaped like a beak. It feeds on corals, seaweeds, sponges and sea anemones. The hawksbill is often hunted for its shell.

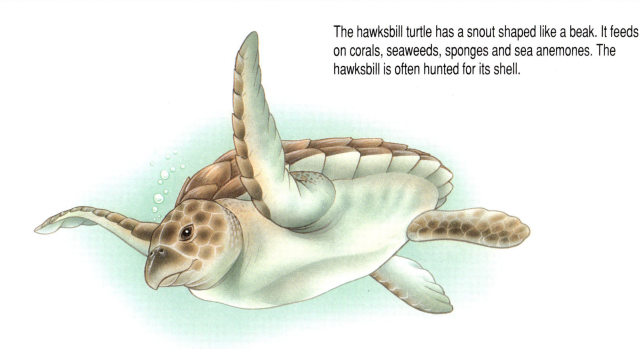

The green turtle is mainly vegetarian and feeds on algae. It nests on tropical beaches around the world.

The female green turtle digs a pit in the sand with her flippers. Then, she lays her eggs in the hole.

When the young turtles hatch, they climb up to the surface of the sand.

The young turtles crawl down the beach to the sea. Many are eaten by predators such as gulls and crabs on their journey.

sea wasp *noun*
A sea wasp is a **jellyfish** with an extremely poisonous sting, which is powerful enough to kill a human in just three minutes, by paralysing the breathing system. The sea wasp has a large, almost transparent bell with four clumps of **tentacles** which are several metres long.
The swimmer did not notice the trailing tentacles of the sea wasp.

sea water *noun*
Sea water is the water in the **sea**. Over hundreds of thousands of years, rivers have flowed into the seas carrying small amounts of salt eroded from the land. Some of the salts are preserved in the sediments at the bottom of the sea. Although water evaporates from the surface of the sea leaving it saltier, it is topped up again by rain water. This cycle keeps the saltiness of most seas the same. Some seas, such as the Dead Sea, are becoming saltier.
Most marine fish can live only in sea water.

sea whip *noun*
A sea whip is a brightly coloured whip-like **coral**, found on muddy parts of the sea-bed. The coral animals, or polyps, bud off from each other along a thin, horny stem, up to 150 cm long. Each polyp has just eight **tentacles**.
The tapering stems of the sea whips were swaying in the water current.

sea-bed ► **ocean floor**

seabird ► page 131

sea-grass *noun*
Sea-grass is a grass-like plant that grows in shallow, warm water on the sandy sea floor and in the **lagoons** behind **coral reefs**. Sea-grass beds act as nurseries for young fish. They are also grazed by animals such as **sea turtles**, dugongs and manatees.
A large green turtle was browsing among the sea-grasses.

seagull ► **gull**

seahorse *noun*
A seahorse is a long, thin fish with a narrow snout. Its body is covered in bony plates, which make it very stiff. Sea dragons are seahorses that look like pieces of seaweed. Female seahorses lay their eggs in a pouch on the male's abdomen. When the eggs hatch, he pushes the young fish out, as though he is giving birth.
Seahorses are so stiff that they cannot bend their body as they swim.

sea-ice *noun*
Sea-ice is ice which is floating on the surface of the **sea**. About one quarter of the ocean is covered in sea-ice at any one time. Some sea-ice forms when the surface of the sea freezes in **polar** regions. **Pack ice**, **ice shelves**, **ice floes** and **icebergs** are also kinds of sea-ice.
During the last ice age, sea-ice reached the coasts of Africa and California, USA.

seal *noun*
A seal is a torpedo-shaped animal that belongs to a group of mammals called **pinnipeds**. There are two kinds of seal. True seals have no ear flaps, and they use their hind flippers for swimming. Sealions and fur seals are eared seals. They have curly ears, and they swim mainly with their front flippers. These flippers can be turned forwards for support on land. Male sealions and fur seals have manes of thick fur, and are much larger than the females.
The diver could see the seals twisting and turning as they hunted fish and squid.

sea-level *noun*
Sea-level is the average height of the surface of the **sea**. The height of dry land is measured at its height above sea-level.
If global warming takes place, many coastal towns will be flooded as the sea-level rises.

sealion ► **seal**

seabird *noun*

A seabird is a bird that lives and feeds at **sea** for most of its life. It has many ways of finding food at sea. **Penguins**, cormorants, **auks**, diving ducks and diving petrels chase fish under water. **Gulls**, storm petrels and **albatrosses** scoop them from the surface of the water. Brown **pelicans**, **gannets**, **boobies** and **terns** dive on their prey from a great height. **Skuas** and **frigatebirds** snatch fish from other birds in mid-air.

The birdwatchers looked for seabirds.

brown booby

Wilson's storm petrel

cormorant

albatross

northern fulmar

pelican

seamount *noun*
A seamount is an underwater mountain formed from an extinct volcano.
Chains of seamounts cross the floor of the Pacific Ocean.

seaweed ► page 133

sediment *noun*
Sediment is the name given to loose particles of solid material which have settled at the bottom of a liquid. **Sand, mud** and the shells of dead **plankton** creatures may settle on the sea floor to form sediment.
The ancient wreck was covered in sediment.

sedimentary rock *noun*
Sedimentary rock is rock that has been formed from sediment. Chalk and limestone are sedimentary rock. They are made from the shells of tiny animals that have settled at the bottom of the ocean over thousands of years. Sandstone is made from sand.
The geologists measured the layers of sedimentary rock in the cliff.

seed shrimp ► **ostracod**

seine net *noun*
A seine net is a net that can be drawn up to form a circle around a shoal of fish. Seine nets may be up to 2,600 metres long and 1,000 metres deep. The most popular seine net is the **purse seine net**.
The fishermen pulled in the seine nets after a day's fishing.

semaphore *noun*
Semaphore is a kind of signalling that uses **flags** or rows of lights to send messages. Different positions of the flags stand for different letters and numbers. Messages sent by semaphore can be sent at a rate of about 25 words a minute. Semaphore is used to signal between ships which are not in radio contact.
The captain used a telescope to read the semaphore signals.

sepia *noun*
Sepia is a purple-brown dye obtained from the ink of **cuttlefish** or **squid**.
The Romans used sepia to colour ink.

Seven Seas *noun*
The Seven Seas today are the **Arctic Ocean, Antarctic Ocean, Indian Ocean,** North and South **Atlantic Oceans,** and the North and South **Pacific Oceans.** Before the 1400s, the Seven Seas mentioned in books were the Adriatic Sea, the Black Sea, the Caspian Sea, the Indian Ocean, the Mediterranean Sea, the Persian Gulf and the Red Sea.
The pirates sailed the Seven Seas in search of gold.

sextant *noun*
A sextant is an instrument in **navigation.** It is used to work out a ship's position on the **sea,** by measuring the angle between a star or the Sun, and the horizon. Once the angle has been measured, the position of the star can be looked up in a very accurate table. This allows the position of the ship to be worked out. Ships can be kept on the right course by taking a series of readings on the sextant.
The captain used the sextant to navigate the English Channel.

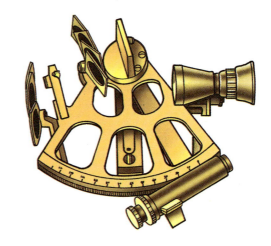

shad ► **herring**

shag ► **cormorant**

seaweed   *noun*

A seaweed is a large alga found in the **sea** and on the seashore. Seaweeds come in many different shapes and sizes. Most of them are stuck to rocks by a tough, sticky pad called a holdfast. Some seaweeds have gas-filled bladders to help them float in the sunlit surface waters. There are three main groups of seaweeds. These are the brown, green and red seaweeds.

Seaweeds provide shelter and food for many small sea creatures.

Most brown seaweeds prefer cold water to live in, where they can grow very large.

sugar kelp

oarweed

thong weed

Red seaweeds are small and delicate and have a feathery appearance. They prefer to live in warm water.

red laver

sea tail

Irish moss

edible dulse

Green seaweeds are small and are able to grow in cold and warm seas.

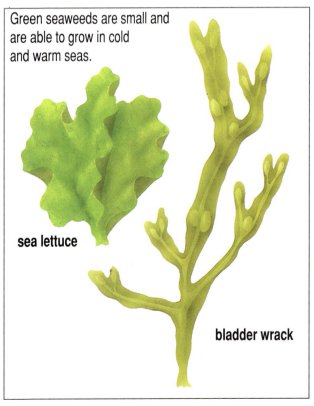

sea lettuce

bladder wrack

shark *noun*
A shark is a **cartilaginous fish**. There are about 350 species of sharks. They live mainly in warm seas throughout the world. Sharks vary greatly in size, from about 13 centimetres to about 12 metres long. All sharks are **carnivores**. They eat live fish including other sharks. Most sharks are also powerful predators. The huge basking shark is a **filter feeder**. It uses strainers on its gills to sieve **plankton** from the water.
Sharks can swim through the water at speeds of up to 50 kilometres per hour.

hammerhead
shark

sharksucker ► remora

shearwater *noun*
A shearwater is a **seabird** which skims the surface of the sea, snatching fish as it flies. Shearwaters are usually dark in colour, with long, narrow wings and a slender bill. They nest in large colonies on cliff-tops and islands. Shearwaters enter and leave their burrows at night.
As night fell, we could hear the shearwaters returning to their burrows.

shellfish *noun*
A shellfish is an animal with a shell that lives under water. **Molluscs**, **crustaceans** and **echinoderms** are all shellfish. Shellfish such as crabs, lobsters, shrimps, mussels and limpets are eaten by people in many parts of the world.
The children looked for shellfish in the rock pools.

shingle *noun*
Shingle is the name given to rounded pebbles which are found in layers on **beaches**. The pebbles are between one and seven centimetres across. They are smooth because they have been worn, or **eroded**, by the **waves**.
It is difficult to walk on loose shingle.

shipping lane *noun*
Shipping lanes are internationally recognized routes taken by ships. Keeping to the right shipping lanes is especially important in narrow channels and **straits**, where many ships are passing in both directions.
The shipping lanes are very close together in the English Channel.

shipworm *noun*
A shipworm is a bivalve **mollusc** that bores holes in the wood of ships, **groynes** and other underwater structures. Shipworms have a worm-like body up to 180 centimetres long. The shell of a shipworm covers only a small part of the front of the animal. It has sharp ridges which cut into the wood. Shipworms line their burrows with calcium carbonate.
The jetty collapsed because it was full of holes bored by shipworms.

shoal ► school

shoal *noun*
A shoal is a bank of **sand**, **mud** or **shingle**. Shoals are found in shallow water in a river or in the sea. They form where a river current slows down and drops the sediment it is carrying.
The ship ran aground on a shoal in the estuary.

shore *noun*
The shore is the land around the edge of the sea or a large lake.
A crowd of people gathered on the shore, waiting for the lifeboat to return.

shorebird ► page 136

shrimp   *noun*
A shrimp is a small **crustacean** with long, slender legs, a fan-shaped tail, and two long antennae, or feelers. Its front legs are armed with pincers. Shrimps feed on small invertebrates and algae. They swim backwards by jet propulsion, jerking their tail under their abdomen. There is no scientific difference between a shrimp and a **prawn**.
Many kinds of shrimps and prawns are sold as food.

silica   *noun*
Silica is a glassy, solid substance composed of the elements silicon and **oxygen**. It forms the transparent shells of **diatoms** and **radiolarians**, and also helps to give strength to the stems of **sea-grasses** and other vegetation.
Under the microscope the students could see the delicate patterns on the silica of the diatom shells.

silt   *noun*
Silt is **sediment** made up of particles that are smaller than sand, usually between 0.02 and 0.06 millimetres across. Silt is often used to describe sediments that build up in places where they create an awkward barrier.
They had to dredge the harbour because it was filling up with silt.

siphonophore   *noun*
A siphonophore is a **cnidarian**. They live together in colonies in which different individuals carry out different tasks. The **Portuguese man-of-war** is an example. The bell is formed by a single huge polyp. Other polyps form prey-catching **tentacles**, feeding structures or reproductive organs.
The Portuguese man-of-war and the Jack-sail-by-wind are siphonophores with large floats that act like sails.

sirenian ► page 138

skimmer   *noun*
A skimmer is a gull-like bird, with a long, blade-like beak. The lower bill is one third longer then the upper bill. Skimmers feed at dusk and at night, by skimming low over the water with the lower bill tip submerged. When the bill touches a **fish** or **crustacean**, the upper bill snaps shut. Skimmers nest on sand banks in colonies of up to 1,000 birds.
A skimmer was flying low over the calm water of the lagoon, creating V-shaped ripples.

skua   *noun*
A skua, or jaeger, is a gull-like **seabird** with a hooked bill, dark plumage and two long central tail feathers. Skuas live as predators and **scavengers** near the nesting colonies of other seabirds. Here, they steal eggs and chicks and feed on dead remains. They also attack other birds in the air to make them drop their food.
The terns rose into the air to drive off the skua that threatened their chicks.

sleeper   *noun*
A sleeper is a **goby** which spends much of its time resting on the sea-bed, as if it was sleeping. Some sleepers hang suspended in the water instead. Unlike most other gobies, sleepers do not have suckers on their abdomen.
As the shark approached, the motionless sleeper dived to the sea-bed to hide.

shorebird *noun*

A shorebird is a bird that lives on the seashore. Shorebirds feed on small **invertebrates** living among the rocks, or buried in the **mud** or **sand**. There are many kinds of shorebird, including knots, godwits, dowitchers, plovers and sandpipers.

Many shorebirds were feeding on the wet mud left by the ebbing tide.

oystercatcher

sandpiper

spoonbill

plover

avocet

curlew

slipper limpet   *noun*
A slipper limpet is a kind of periwinkle which is unusual because it can change sex. Slipper limpets live in piles on top of one another. All slipper limpets start life as males. When a male settles on a rock, it changes into a female. A male limpet then settles on top of the new female and mates. But as soon as another male settles on top of it, it too changes into a female.
Male slipper limpets were at the top of the pile, and females at the bottom.

smelt   *noun*
Smelts are a group of small, silvery fish, which are very similar to salmon. Salt water smelts live in The Atlantic, Pacific and Arctic Oceans. There are about 10 species of smelts. They usually measure less than 20 centimetres in length. Smelts are easy to catch when they move up-river to **spawn**. They are a valuable food fish. Smelts have very fatty flesh, full of natural oils. American Indians used them as natural candles, and called them candlefish.
A capelin is a member of the smelt family.

snapper   *noun*
A snapper is a tropical fish with a large mouth and pointed pectoral fins. A snapper also has a large mouth with strong teeth. It can be red or greenish in colour, or striped. Snappers usually have a forked tail and large scales, with a well-marked line along them. They can reach 60 to 90 centimetres in length.
Snappers are often brightly coloured, and many are good to eat.

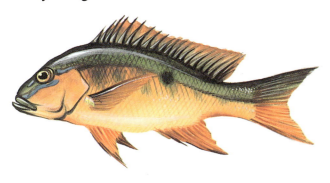

snipe eel   *noun*
A snipe eel is a curious **deep sea fish**, with large eyes and an extremely long, slender, tapering body. Its jaws are often very long, and curve away from each other, so that they do not meet except near the head. Snipe eels feed on **crustaceans**, but it is not known why they have such strange jaws.
When the fishermen brought up the nets from deep water, they found a snipe eel caught by its jaws.

snorkel   *noun*
A snorkel is a J-shaped, short rubber tube with a mouthpiece attached to the curved end. It allows a swimmer to breathe while looking down into the water. The mouthpiece is gripped between the teeth and the swimmer can then put their head under water.
The swimmer used a snorkel so that she could breathe below the surface of the sea.

sodium chloride ► **salt**

soft coral   *noun*
A soft coral is a coral that does not have a hard skeleton. Instead, it has small fragments of calcium carbonate scattered throughout its fleshy body. Soft corals form branching, feathery colonies which sway in the current. They are often very brightly coloured.
The large cavern in the reef was almost hidden by a curtain of soft corals.

sole ► **flatfish**

sonar   *noun*
Sonar is a system which uses echoes, or reflected sound waves, to detect objects under water or to measure the **depth** of the sea floor. An instrument sends out pulses of sound which bounce back from solid objects, the sea floor or shoals of fish.
Echo-sounding is a form of sonar.
Oceanographers use sonar to chart the sea-bed.

sirenian *noun*

A sirenian is a slow-moving marine mammal that feeds on plants. Sirenians are divided into two families. These are dugongs and manatees. There are three **species** of manatee, which are all found in rivers, estuaries and shallow coastal waters. The dugong is a marine species which is found in the warm, shallow coastal waters of the west Pacific and Indian oceans.

The marine biologist was studying sirenians.

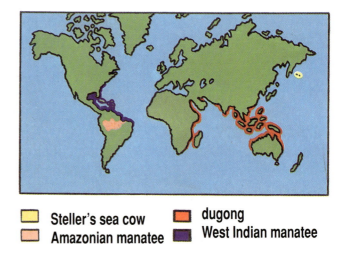

- Steller's sea cow
- Amazonian manatee
- dugong
- West Indian manatee

Manatees mate in the autumn. The female gives birth about 10 months later. The calf feeds from a nipple which is found under the mother's flippers.

A manatee is about five metres long and weighs about 650 kilograms. It feeds on water plants and can eat more than 45 kilograms of them in a day.

The dugong is a plant-eating mammal. It has a blunt, rounded snout and a bristly upper lip. It has a notched tail, and uses its flipper to swim and to push sea grass near its mouth. Dugongs can be up to three metres long and weigh 300 kilograms.

A giant sirenian, known as Steller's sea cow, was discovered in 1748. Steller's sea cows were widely hunted for their hide and, by 1768, they were extinct.

South Equatorial Current ▶
Equatorial Current

Southern Ocean ▶ Antarctic Ocean

spawn *verb*
To spawn means to lay eggs in water. When fish spawn, the female lays her eggs and the male then sheds his milt over them.
The salmon were spawning in the shallow water.

spawn *noun*
Spawn is the name given to the eggs laid by **fish, molluscs**, amphibians and other animals. The eggs are usually produced in large quantities because many of them are eaten by aquatic animals.
Spawn, or roe, is often used as a food product.

spawning ground *noun*
A spawning ground is a special place where fish gather to **spawn**. Many fish **migrate** over long distances to spawning grounds. Salmon and smelts travel from the ocean to rivers to spawn.
The Sargasso Sea is the spawning ground of the European eel.

spear-fishing *noun*
Spear-fishing describes the use of a spear to catch fish. Spears are sometimes used to catch fish and octopus in shallow water. They are usually used by divers under water.
Spear-fishing is a popular sport in the warm waters off the Bahamas.

species *noun*
Species is a term used to group together plants and animals of the same kind. The sperm whale is a species of whale. The herring gull is a species of seagull. Animals, plants and other **organisms** only mate with members of their own species.
There are about 21,000 species of fish.

spermaceti *noun*
Spermaceti is a wax which is found in the head of **sperm whales** and other **cetaceans**. The wax is solid at room temperature, but melts easily. As it melts it changes **density**, and some scientists think it may help the **whale** to adjust its **buoyancy**, or it may be involved in producing the sounds the animal makes during **echolocation**.
The head of a sperm whale may contain one tonne of spermaceti.

sperm whale *noun*
The sperm whale is a large **whale** with a huge, almost square head and a narrow lower jaw. The mouth is on the underside of the whale's snout, and is armed with large, cone-shaped teeth. Male sperm whales grow up to 19 metres long. Females are much smaller. Sperm whales are good divers and are thought to reach depths of over 3,000 metres. They feed mainly on **squid**.
Sperm whales use echolocation to find their prey in the dark ocean depths.

spiny lobster *noun*

A spiny lobster is a **lobster** which has no large claws or pincers, but has spines on its shell and feelers or antennae. They hide by day in cracks in the rocks or among corals, and come out to hunt at night. Many spiny lobsters **migrate** at certain seasons. They move in long lines, keeping in touch using their antennae.

In autumn, spiny lobsters off the Florida coast migrate to deeper water to avoid the effects of winter storms.

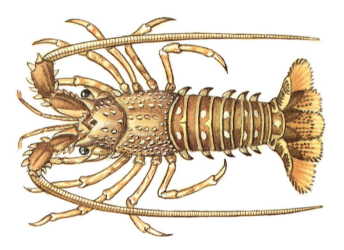

spiracle *noun*

A spiracle is a round hole leading from a fish's gill chamber to the outside of its body. Spiracles are only found in **cartilaginous fish** and **jawless fish**. Bony fish do not have a spiracle. Instead they have a flap-like **operculum**.

Water enters through the lamprey's mouth, passes over its gills and leaves through the spiracles.

spit *noun*

A spit is a narrow ridge of **sand** or **shingle** which is joined to the land at one end of its length. Spits often form across the mouths of **estuaries**, **harbours** and shallow **bays**. They are formed from **sediment** which has been moved along the **beach** by **longshore drift**. Many young fish live in sheltered lagoons behind sand spits.

They saw that the spit separated the sea from the river.

sponge *noun*

A sponge is a simple animal which feeds on **microscopic**-sized **organisms** in the water. The sponge draws a stream of water through its body by beating thousands of tiny hairs called **flagella**. Water enters through holes in the sides of the sponge, and leaves through a larger hole at the top. Sponges are fixed to the sea-bed, or to rocks or **corals**. Some are vase-shaped or tree-like. Others form large, flat crusts.

On the floor of the Antarctic Ocean live large sponges which are many hundreds of years old.

spookfish *noun*

A spookfish, or barrel-eyes, is a small, dark brown, **deep sea fish** whose eyes are at the top of upward-facing tubes. Spookfish detect the silhouettes of their **prey** against the light coming from above. They confuse their own **predators** by producing light, which they beam downwards from their flattened abdomens, which act like reflectors.

Spookfish can only look upwards, when they are hunting for prey.

spoonbill *noun*

A spoonbill is a large, long-legged bird, with a very large spoon-shaped bill. It feeds in shallow water, swinging its bill from side to side to catch small **molluscs**, **crustaceans**, **fish** and **worms**. Its bill is very sensitive to touch and vibrations.

Large numbers of spoonbills, herons and ibises feed in the mangrove swamps.

spring tide   *noun*
A spring tide is a tide with a large difference between the high water mark and the low water mark. It takes place twice a month at full Moon and new Moon, when the gravities of the Sun and the Moon pull in the same direction. The opposite of a spring tide is a **neap tide**.
Driven by the strong wind, the spring tide burst through the sea defences.

squid   *noun*
A squid is a **cephalopod**. It has a tapering, cylinder-shaped body and a long, narrow **fin** on each side. Squids have no bones. Instead, they have a horny shell or pen, inside their body to provide support. Squids have 10 **tentacles**, two of which are very long and thin with rows of suckers at their tips. These are used to catch **prey**. Squid swim in large shoals, by **jet propulsion**. If they are alarmed, they squirt out a cloud of ink to confuse their attacker.
Most squids are between 30 centimetres and almost 12 metres in length.

stack   *noun*
A stack is a large pillar of rock which stands alone at the base of a cliff. It has been formed by the power of waves which have **eroded** a cave out of the cliff edge. The cave has been eroded into an arch, and later into a stack.
Many seabirds can be found nesting on stacks on the coast.

stack

starboard   *noun*
The starboard side is on the right hand side of the front of a ship. The opposite of starboard is **port**.
The sea turtle swam to starboard of the ship.

starfish   *noun*
A starfish is a star-shaped **echinoderm**. Most **species** of starfish have five arms attached to a central disc, but some have more. The mouth is on the underside of the disc. Starfish live on the sea-bed, on **coral reefs** and in rock pools. They feed mainly on **molluscs** and **crustaceans**. The suckers on the tube feet of starfish are strong enough to pull apart the shells of mussels.
A large starfish was creeping over the sea-bed towards the mussels.

stargazer   *noun*
A stargazer is a **fish** that lives on the sea-bed. It has a large, flat head with a big mouth which slants upwards, and eyes on top of its head. Some stargazers have electric organs behind their eyes, which warn of approaching fish and may perhaps also be used to stun their **prey** and anyone who steps on them, with electric shocks. They also have two poisonous spines above each pectoral **fin**.
Some kinds of stargazer produce enough poison to kill a human.

stern   *noun*
The stern is the back of a ship. The opposite of stern is **bow**.
The ship's mate was in the stern of the ship, with his hand on the rudder.

stickleback   *noun*
A stickleback is a small **fish** which has spines on its back which can be raised to defend itself against attackers. The stickleback can stick its spines in a **predator's** throat, so that the stickleback is spat out and survives. Sticklebacks are found in fresh water and along **coasts**.
The male stickleback nests in algae.

stingray *noun*
A stingray is a **ray** with a long, tapering, flexible tail, armed with one or more poisonous spines. It lives buried in **sand** or **mud** on the beds of rivers, lakes or the **sea**. Stingrays feed on **shellfish** and other small **invertebrates**. If a stingray is stepped on, it will lash its tail so that its spines are pushed into the foot or leg. The spines cause serious wounds which are extremely painful. They can be fatal if not treated.
The flat-topped teeth of stingrays are adapted for crushing shellfish, so they can be eaten.

stomotopod ► **mantis shrimp**

stonefish ► **scorpionfish family**

storm *noun*
A storm is a strong wind.
A storm measures force 10 or 11 on the Beaufort Scale.

storm surge *noun*
Storm surge describes large **waves** that appear without warning, either out at sea or on the coast. They are caused by a storm far out at sea. Storm winds stir up big waves which roll across the ocean until they reach the coast, several days later.
News of the storm surge brought many surfers to the beach.

strait *noun*
A strait is a narrow strip of sea that links two seas or oceans.
Pirates lie in wait for ships in the Strait of Malacca, between the Indian Ocean and the Pacific Ocean.

Strait of Gibraltar *noun*
The Strait of Gibraltar lies between the southern tip of Spain and the coast of north-west Africa. It connects the **Mediterranean Sea** with the **Atlantic Ocean**.
The narrowest part of the Strait of Gibraltar is only 22.5 kilometres wide.

Strait of Hormuz *noun*
The Strait of Hormuz lies between the south coast of Iran and the northern tip of Oman, on the Arabian Peninsula. It links the **Persian Gulf** and the **Arabian Sea**.
Many tankers sail through the Strait of Hormuz.

Strait of Magellan *noun*
The Strait of Magellan lies between the southern tip of the South American mainland and the island of Tierra del Fuego in Chile. It links the Atlantic and Pacific oceans.
The ship had a stormy passage through the Strait of Magellan.

strandline *noun*
A strandline is a line of material washed up, or stranded, by the tide. It marks the highest point on the beach which was reached by the last **high tide**. The strandline may contain clumps of **seaweed**, **mollusc** shells, the cast-off shells of **crabs**, other animal remains and human rubbish.
It is always exciting to explore the strandline after a storm.

stromatolite *noun*
A stromatolite is a blue-green **alga**. It forms mounds of crusty layers made of trapped **sediment** and calcium carbonate, or limestone. Stromatolites live near the low water mark in warm, shallow seas where there is a high level of evaporation. Stromatolites were some of the first forms of life to appear on Earth.
There are fossils of stromatolites in rocks which are more than 600 million years old.

sturgeon *noun*

A sturgeon is a large fish which is covered in rows of bony plates. Its tail is forked. Sturgeons have a long snout. They use **barbels** on their snout to feel for small fish and **invertebrates** in the **mud**. Most sturgeons live in the sea but **migrate** to rivers to **spawn**. Sturgeons are caught for their flesh and for their eggs, which are called caviar.
Some sturgeons grow up to three metres long, and may be up to 300 years old.

subduction zone *noun*

A subduction zone is an area on the Earth's crust where one tectonic plate has been forced down into the mantle underneath a continental plate. This happens because the tectonic plates are denser than the granite continental plates they are forced under. This has created a deep trench on the **ocean floor**. Subduction zones lie below oceanic trenches.
Part of the ocean floor disappears in a subduction zone.

sublittoral zone *noun*

The sublittoral zone is part of the sea-shore. It is found just below the low water mark. The sublittoral zone is also the area of the ocean that stretches from low tide to the edge of the continental shelf.
The divers were looking for octopus as they swam in the sublittoral zone.

submarine *adjective*

Submarine describes an object or habitat which is found below the surface of the ocean.
Not far from the shore are submarine forests of kelp.

submarine *noun*

A submarine is a ship which can travel under water as well as on the surface. It contains tanks that are filled with water to make it sink. Submarines range in size from about 60 metres to over 150 metres in length. The hull is usually about nine metres across. About 100 crew members live and work aboard submarines. Most submarines are used as warships.
Some submarines can stay under water for many months.

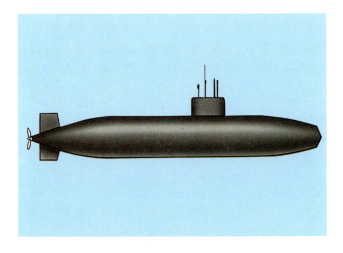

submarine cable *noun*

A submarine cable is a line laid on the **ocean floor** which carries telephone and telegraph signals. Submarine cables are well-protected and flexible to help them cope with water movements. The cables are laid by a special cableship. Radio and satellite signals have been invented since the first submarine cables were laid, but cables are still cheaper to use and more confidential.
The world's longest submarine cable runs for 15,151 kilometres between Canada, Fiji, Australia and New Zealand.

submarine canyon *noun*

A submarine canyon is a deep gorge cut into the **continental slope**. It may have been cut by a river at a time when the sea-level was lower, or it may have been formed by **turbidity currents**.
The Grand Bahama submarine canyon is almost five kilometres deep.

submersible   *noun*
A submersible is a small underwater vessel which is used for studying under the **sea**. Submersibles usually carry a small crew of two or three people. Submersibles have electrically-powered propellers which allow them to be moved in many different directions. They have remote controlled tools which can be operated by the crew. These are used to collect samples from the sea-bed.
Submersibles can go down to depths of at least 3,050 metres.

substrate   *noun*
Substrate describes a substance on which an object rests or feeds, or the surface to which an object is attached.
Seaweeds are attached to the substrate by holdfasts.

Subtropical Convergence   *noun*
A Subtropical Convergence can occur anywhere just north and south of the tropics where **ocean currents** meet. Warm surface water driven towards the poles by the **Trade Winds** meets colder water driven towards the **equator** by the **westerlies**.
At the Subtropical Convergence, water is forced down into the ocean depths.

Suez Canal   *noun*
The Suez Canal is a canal in Egypt. It connects the northern-most tip of the **Red Sea** with the **Mediterranean Sea**. This makes it possible for ships to travel from the Mediterranean Sea to the **Indian Ocean**, without having to sail around Africa.
Oil tankers use the Suez Canal on their way from the Persian Gulf to Europe.

sulphur bacteria   *noun*
Sulphur bacteria are **bacteria** that live around hot springs which produce sulphur minerals. They use the sulphur compounds to produce energy.
The food chain of hydrothermal vents depends on sulphur bacteria.

Sundarbans   *noun*
The Sundarbans is the world's largest **mangrove swamp**. It covers parts of the **delta** of the Ganges, Hooghly and Meghna Rivers in Bangladesh, India.
Crocodiles lurk among the mangroves of the Sundarbans.

sunfish   *noun*
Sunfish is the name given to several kinds of fish. The most common type of sunfish is the ocean sunfish. It has an almost circular body and large dorsal and anal fins which are almost opposite each other. Ocean sunfish can grow up to four metres long, but the average size is about 90 centimetres. They are often found in the open ocean, sometimes at depths of 350 metres. Ocean sunfish feed on **jellyfish**, **sea gooseberries** and young fish.
Sunfish can often be found living in fresh water.

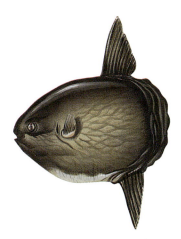

surf   *noun*
Surf is a mass of foam which is caused by **waves** breaking on the **shore**, or **reef**, or similar area.
Whilst walking on the beach they saw some seals playing in the surf.

surface current   *noun*
A surface current is an ocean current found in the upper 100 metres of the sea. It is caused by the wind.
Warm surface currents carried the sea turtles towards their breeding beaches.

surface tension *noun*
Surface tension describes how the surface of a liquid acts like a sheet of elastic. In water, the water particles attract each other, forming a tight skin and holding the surface layers together. Small **organisms** can rest on the surface of the water without sinking.
Drops of water fall in a round shape because of surface tension.

surfing *noun*
Surfing is a sport in which people ride waves breaking towards the shore on special surf-boards. It is also possible to surf without a board by holding the body very stiff.
There is often good surfing on the California coast.
surf *verb*

surge ► **storm surge**

surgeonfish *noun*
Surgeonfish are a group of **coral reef fish** which have very sharp spines on either side of the tail. These spines lie in grooves, but they can be flicked up to slash at enemies if the fish is disturbed. Surgeonfish feed on small invertebrates and algae which they scrape from rocks and **reefs**. There are about 100 species of surgeonfish. The surgeonfish family includes tangs, unicornfish and the **Moorish idol**.
Surgeonfish usually travel in small groups.
The bright colour of the surgeonfish's spines is a warning to predators.

survey ship *noun*
A survey ship is a ship used in the study of the sea-bed. It has **sonar** equipment on board to record the sea-bed on a graph.
They used the sonar equipment on board the survey ship.

suspension feeder *noun*
A suspension feeder is an animal that feeds on particles of food which are suspended in the water. They may filter out the particles using feathery arms or sticky mucus, or they may draw the water through filters inside their bodies. Cockles, mussels, **barnacles**, **feather stars** and **radiolarians** are suspension feeders.
Many marine worms are suspension feeders.

swallower *noun*
A swallower is a dark-coloured **deep sea fish**, with a long tapering body and an enormous mouth. It has a large, elastic stomach which can stretch to swallow **fish** up to twice its own size. Some swallowers are almost two metres long.
The whip-like tails of swallowers have a light at the tip.

swell *noun*
Swell describes the arrival of large **waves** in an area of fairly calm water. Swells are caused by winds blowing some distance away.
The swell took the swimmers by surprise.

146

swim bladder   *noun*

A swim bladder is a gas-filled bag found inside the body of a fish which helps the fish to float. The fish can alter the amount of gas in its swim bladder depending on whether it wants to rise or sink in the water.
Sharks have no swim bladders, so they sink if they stop swimming.

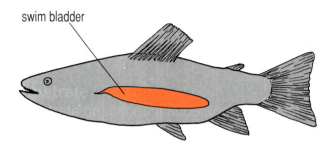
swim bladder

swordfish   *noun*

A swordfish has a long, narrow, pointed snout, which has many small teeth. The snout looks like the blade of a saw.
A swordfish probably uses its 'sword' to slash at prey. Swordfish are closely related to marlins.
The large hole in the side of the boat was made by a swordfish.

symbiosis   *noun*

Symbiosis describes a partnership between two different kinds of **organism** which helps both organisms to survive. For example, some hermit crabs place sea anemones on their shells to protect them from their enemies with its stinging tentacles. In return, the anemones are taken to new feeding areas. They also feed on scraps dropped by the crab.
The behaviour of cleanerfish is an example of symbiosis.

tankers   *noun*

A tanker is a ship which is designed to carry large amounts of liquid cargo. Most tankers carry **petroleum** or petroleum products, but some carry chemicals, wine, coal, grain or iron ore.
The oil tanker was carrying crude oil.

tarpon   *noun*

A tarpon is a large **fish** with big, shiny **scales**. Some tarpons grow up to 2.5 metres long. The last ray of a tarpon's dorsal **fin** is very long. Tarpons have teeth on the roof of the mouth and on the tongue as well as on the jaws. They are fast swimmers, and are famous for making huge leaps out of the water.
The old angler enjoyed a good fight with a leaping tarpon.

Tasman Sea   *noun*

The Tasman Sea is part of the South **Pacific Ocean**. It is found between south-east Australia and the west coast of New Zealand. The Tasman Sea covers about 2,300,000 square kilometres.
The fishing boat was caught in a storm on the Tasman Sea.

tectonic plate   *noun*

A tectonic plate is a separate piece of the Earth's **crust** which is made up of seven major tectonic plates and several smaller ones. The theory of **plate tectonics** describes how tectonic plates move around the surface of the Earth.
Earthquakes and volcanoes occur where tectonic plates collide.

teleost *noun*
A teleost is a bony fish which has a tail **fin** divided into two equal parts, and a gas-filled **swim bladder**.
Most fish are teleosts.

temperature *noun*
Temperature measures how hot or cold something is. Temperature is measured in degrees, using a thermometer. The temperature of the ocean varies with the seasons and with the direction of the **prevailing wind**.
The temperature of the ocean becomes less with depth, because the deep water is further from the Sun's rays.

temperature profile *noun*
The temperature profile of the **ocean** describes how its temperature changes with **depth**. This depends on how much the **wind** mixes the surface waters. If there is a lot of mixing, the water may be warm to a depth of several hundred metres. The still waters below this level steadily decrease in temperature. If there is little wind, the temperature steadily drops from the surface to the **ocean floor**.
The oceanographers measured the temperature profile of the ocean.

tentacle *noun*
A tentacle is a long, slender, flexible structure used by animals for feeling or gripping objects, or for moving. **Jellyfish** use stinging tentacles to catch their prey. Some fish have tentacles, called **barbels**, below their mouth for feeling and tasting.
A squid's tentacles are armed with suckers.

tern *noun*
A tern is a slender **seabird** with long wings and a forked tail. Terns are sometimes called 'sea swallows'. Most terns have a long, slender, pointed bill. They feed on fish and small **invertebrates**, by hovering over the water, then swooping on their prey.
The terns were feeding fish to their chicks.

thermocline *noun*
Thermocline describes the **depth** in a lake or **sea** below which the **wind** no longer mixes the water. The temperature may fall sharply below the thermocline.
The autumn storms upset the thermocline.

tidal *adjective*
Tidal describes something which is affected by the **tides**, or which is caused by the ebb and flow of the tides.
Tidal also describes something which is affected by the gravities of the Sun, Moon and Earth.

tidal current *noun*
A tidal current is a horizontal movement of water caused by the effects of **gravity** from the Sun, the Moon and the Earth.
The boat was caught in a tidal current.

tidal power *noun*
Tidal power is electricity produced by electric generators which are driven by the ebb and flow of the **tides**. Tidal power stations use **barrages** to trap the water.
Very few power stations use tidal power.

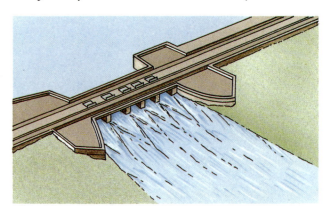

tidal range *noun*
Tidal range describes the difference between high and low **tides**. Tidal range is measured as the difference in water **depth** at a given place between low water and high water.
The Severn Estuary in Great Britain, is a very good example of an estuary with a high tidal range.

tidal wave *noun*
A tidal wave, or tsunami, is a giant wave. It may be over 30 metres high. A tidal wave is caused by movements of the **ocean floor**, due to **earthquakes** or volcanic eruptions. A tidal wave has nothing to do with the tides although sometimes the **sea** retreats from the **shore** before the wave arrives. A tidal wave is often followed by a series of smaller **waves**.
After the eruption of Krakatau volcano in 1883, tidal waves killed 36,000 people.

tide ► page 150

tide pool *noun*
A tide pool is a pool of water left behind by the ebbing tide.
Many small animals live in tide pools.

toadfish *noun*
A toadfish is a toad-like **fish** which hides among **seaweeds** or in the **mud** of the sea-bed, where it lies in wait for **prey**. Toadfish have a toad-like head, with bulging eyes and a large mouth. They prop themselves up on their pelvic **fins**.
The croaking sound was made by a toadfish.

Trade Winds *plural noun*
The Trade Winds are **prevailing winds** that blow towards the **Equator** in the **tropics**. They blow from the north-east in the **northern hemisphere** and from the south-east in the **southern hemisphere**.
The Trade Winds bring clear, sunny weather to ocean islands.

trawling *noun*
Trawling is a method of fishing. In trawling, a cone-shaped **net**, or trawl, is dragged through the water or along the sea-bed, behind a boat. A boat that pulls a trawl is called a trawler. Trawling is used to catch **shrimps**, **herring** and bottom-dwelling fish such as **cod**.
The fishermen were trawling for cod.

trench *noun*
A trench is a long, deep, steep-sided valley, found in the deep **ocean floor**. Trenches form where one **tectonic plate** bearing part of the ocean floor is pushed underneath another plate.
Some ocean trenches are 11 kilometres deep.

tributyltin (TBT) *noun*
Tributyltin, or TBT, is a chemical found in paints. It helps to prevent **fouling** on the bottom of boats. It is extremely poisonous to marine life, even in very small amounts.
The use of TBT is banned in Britain.

triggerfish *noun*
A triggerfish is a deep-bodied **fish** which has a high spine at the front of the first dorsal **fin**. This spine can be raised and locked in position like a trigger, by a second spine. The trigger fish uses this spine to jam itself in crevices. Most triggerfish are less than 45 centimetres long.
As the shark approached, the triggerfish wedged itself among the corals.

tripodfish *noun*
A tripodfish is a fish that lives on the deep sea floor. Its pelvic **fins** and the lower part of its tail fin are long and stiff. They form a tripod to prop up the fish on sea-bed. This helps the tripodfish to search for food above the **mud**.
The tripodfish was walking over the sea-bed on its stiff fins.

Tropic of Cancer ► **tropics**

149

tide *noun*

A tide is the regular movement of sea water towards and away from the land. Tides are caused by the gravities of the Moon and Sun, which pull the sea away from the Earth. Tides with the smallest difference between high and low water are called **neap tides**. Tides with the largest difference between high and low water are called **spring tides**. *When the children visited the beach they could see that the tide was in.*

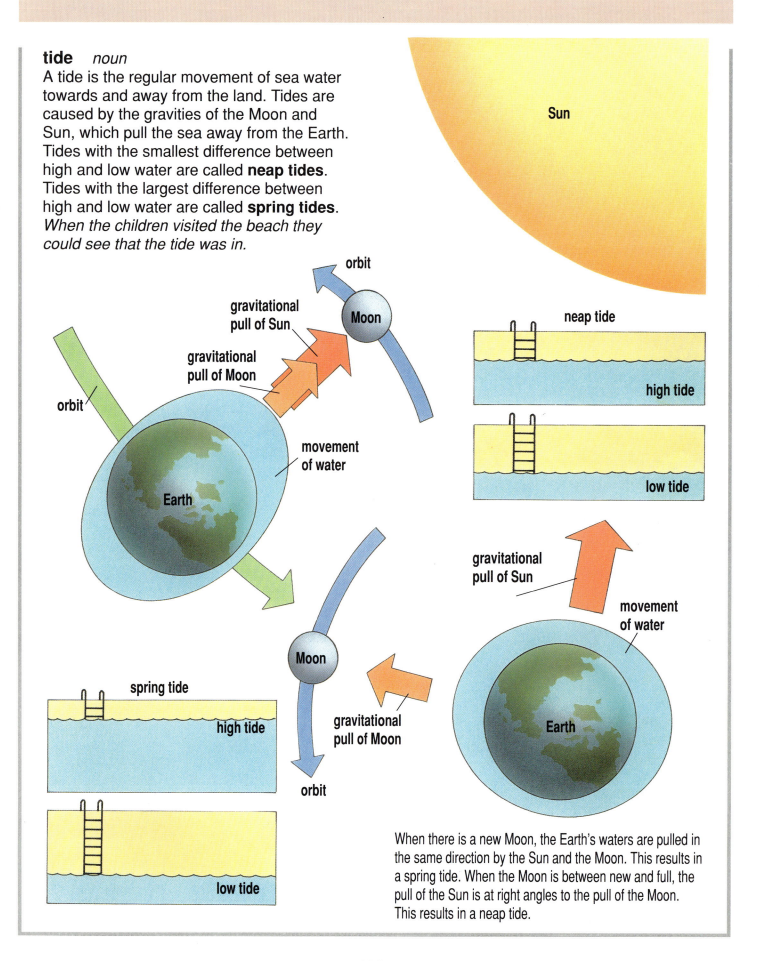

Sun

orbit

gravitational pull of Sun

gravitational pull of Moon

Moon

orbit

movement of water

Earth

Moon

spring tide

high tide

low tide

gravitational pull of Moon

orbit

neap tide

high tide

low tide

gravitational pull of Sun

movement of water

Earth

When there is a new Moon, the Earth's waters are pulled in the same direction by the Sun and the Moon. This results in a spring tide. When the Moon is between new and full, the pull of the Sun is at right angles to the pull of the Moon. This results in a neap tide.

Tropic of Capricorn ► tropics

tropical   *adjective*
Tropical describes something that is found in the **tropics**.
Porcupinefish live only in warm tropical waters.

tropics   *noun*
The tropics describes part of the Earth's surface, found between **latitude** 23.5 degrees north and latitude 23.5 degrees south. Latitude 23.5 north is called the Tropic of Cancer. Latitude 23.5 degrees south is called the Tropic of Capricorn. Most places in the tropics have warm or hot temperatures throughout the year. Tropical climates do not have a cool season and often have heavy rainfall.
Warm water from the tropics is carried towards the poles by ocean currents.

trumpetfish   *noun*
A trumpetfish is a long, slender **fish**, with a stiff body and a long, tubular snout that flares out at the tip, rather like a trumpet. Trumpetfish drift along in the water, using their transparent **fins**.
Trumpetfish are usually dull in colour and can camouflage themselves very well to escape from predators.

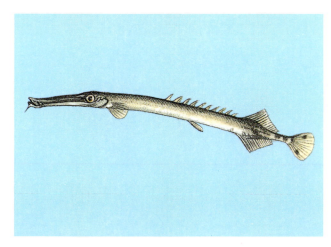

trunkfish ► **boxfish**

tsunami ► **tidal wave**

tube worm   *noun*
A tube worm is a kind of marine **worm** which builds hard tubes around itself for protection. Sand mason worms and peacock worms build tubes of **sand** grains cemented together with **mucus**. Worms such as sabellaria build large reefs of sand grain tubes, joined together like honeycombs. Serpulid worms make stiff white tubes of **calcium carbonate**. Some of these are upright, while others are coiled, and look like snail shells.
A colourful fan of feathery tentacles emerged from the tube as the peacock worm began to feed.

tugboat   *noun*
A tugboat, or tug, is a small boat which is used to tow or push barges and ships. Tugboats are used to help large ships dock in small **harbours**, and to **salvage** wrecks. Tugs usually measure about 20 to 75 metres long.
Two tugboats helped the liner to reach the dock.

tuna ► **mackerels and tunas**

tunicate ► **sea squirt**

turbid   *adjective*
Turbid describes water that is stirred up and muddy. Turbid water is often caused by storms and bad weather.
It was difficult to see through the turbid water.
turbidity   *noun*

turbidity current　*noun*
A turbidity current is a water movement which sweeps **sediment** from the **continental shelf** or **continental slope** down to the **abyssal plain**. A turbidity current looks like a puffy, brown cloud. The water is so full of **sediment** particles that it moves like a thick liquid.
Powerful turbidity currents have carved deep canyons in the continental slopes.

turbot ► flatfish

turbulence　*noun*
Turbulence is an uneven movement of water or air. It happens when water or air flows rapidly past an object, or when **wind** blows over the surface of water.
During the storm, the turbulence spread from the surface waters to the sea-bed.
turbulent　*adjective*

turnstone　*noun*
A turnstone is a **shorebird** with a short, flattened beak which is turned up at the end, which it uses to turn over stones and shells in search of food.
In summer, turnstones migrate to the Arctic tundra to breed.

tusk shell　*noun*
A tusk shell, or tooth shell, is a marine **mollusc** with a tubular, tusk-shaped shell which is open at both ends. Tusk shells live buried in **mud** or **sand**. They use **tentacles** to catch tiny water creatures, and a muscular foot to bury themselves in the sand or mud.
Indians once used tusk shells as money.

typhoon ► cyclone

Tyrrhenian Sea　*noun*
The Tyrrhenian Sea is part of the **Mediterranean Sea**. It lies between the islands of Corsica, Sardinia, Sicily and the west coast of the Italian mainland.
The island of Elba lies in the Tyrrhenian Sea.

undertow　*noun*
An undertow is a strong flow of water that moves away from the **shore** along the sea-bed. It takes the water from the breaking **waves** back out to **sea**.
She was swept off her feet in the deep water by the undertow.

upwelling　*noun*
Upwelling describes an area where cold, deep **ocean** water rises to the surface. This happens along **coasts** where the **prevailing wind** blows **offshore**. The **surface currents** move water away from the shore, and deep water wells up to take its place. This deep water is rich in **nutrients** where dead **organisms** have decayed on the **ocean floor**.
The anchovy fishery off the coast of Peru depends on a nutrient-rich upwelling.

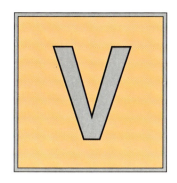

velocity profile   *noun*
A velocity profile is a kind of measurement. It shows the speed at which water moves horizontally at different depths in the ocean. A velocity profile also shows the **depth** of various **ocean currents**.
From the velocity profile, the scientists saw that the surface current was only 100 metres deep.

vertebrate   *noun*
A vertebrate is an animal with a backbone. There are about 40,000 species of vertebrates, including **mammals**, birds, reptiles, amphibians and **fish**. Animals without backbones are called **invertebrates**.
Whales, sea turtles and tunas are marine vertebrates.

vertical migration   *noun*
A vertical migration is the regular movement of **marine** creatures from one water level to another in the **ocean**. Many of the creatures that make up **plankton** move away from the surface by day, and return to the surface at night.
Fish predators follow plankton on their vertical migrations.

viperfish   *noun*
A viperfish is a small **deep sea fish**, with large, needle-like, backward-curving teeth, rather like those of a viper. It has **light organs** along its side and abdomen, and sometimes also on the tips of its **fins** and inside its mouth.
The light organs inside the mouths of viperfish lure other fish to their deaths.

viscosity   *noun*
Viscosity describes how easily a liquid flows. Syrup has a greater viscosity than water because it flows more slowly and does not spread out so far over a flat surface.
Fresh water has a lower viscosity than salt water.
viscous   *adjective*

viviparous fish   *noun*
A viviparous fish is a **fish** which gives birth to fully-formed young fish. Many **sharks** are viviparous.
Viviparous fish do not produce so many young as egg-laying fish.

volcanic island   *noun*
A volcanic island is the top of an underwater **volcano** which shows above the surface of the **sea**. The volcano may still be active, or it may be extinct.
The Hawaiian Islands are volcanic islands.

volcano   *noun*
A volcano is a hill or mountain formed from **lava** and other material that has forced its way upwards from below the Earth's crust. From time to time a volcano may throw out red-hot **magma**. Gas trapped in the magma may explode, throwing out pieces of rock, dust and steam. When the **tectonic plate** bearing the volcano moves away from the **hot spot** in the **mantle** below, the volcano becomes extinct, and no longer erupts.
There are chains of old volcanos on the floor of the Pacific Ocean.
volcanic   *adjective*

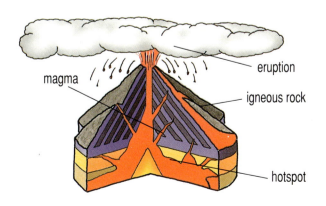

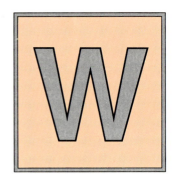

wader ► shorebird

walrus *noun*
A walrus belongs to a group of **marine mammals** called **pinnipeds**. It is a large animal, which can be up to 3.7 metres long. Two thick tusks up to a metre long curve down from its upper jaw. The walrus uses these tusks to lever clams from the sea-bed, and to haul itself out of the sea onto the **sea-ice**. Walruses have greyish skin covered in thin, red hair. They live in large groups.
The beach was covered in dozing walruses.

water cycle *noun*
The water cycle is the process by which water leaves the Earth's surface and returns to it again. Water **evaporates** from the surface of **seas**, lakes and rivers and enters the atmosphere as water vapour. As the vapour cools high in the atmosphere, it turns into water droplets, which in turn form clouds. These drop rain on the Earth's surface. The water drains through the ground into rivers and back to the sea.
Thanks to the water cycle, there is enough water for life to survive on land.

waterspout *noun*
A waterspout is a whirling funnel of water. It rises from the surface of the **sea** or lake, and ends up in a large cloud. It sucks up water droplets into the cloud. A waterspout is like a small, intense **cyclone**. Most waterspouts occur in **tropical** regions.
The sailors watched the waterspout moving across the sea.

wave *noun*
A wave in the **sea** is like a ridge of water that moves along the surface. The water does not move far. Each water **particle** moves up and forwards, then down and back, in a circle. Waves are caused by the **wind** pushing against the water surface. When a wave reaches the **coast**, the sea-bed is shallower than the bottom of the wave. The wave becomes higher and the tops of the ridges, or crests, get closer together, until they tumble over, or break, to form **breakers**.
We heard the waves crashing onto the shore.

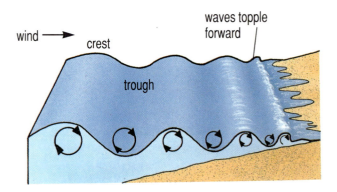

wave energy ► wave power

wave power *noun*
Wave power is energy obtained from **waves**. The up-and-down motion of the waves can be used to drive an electric generator, using machines like the **Salter's Duck**. Energy from the waves has been harnessed at the La Rance tidal barrage in Brittany, France. This kind of energy is also known as tidal power.
Wave power is a very clean source of electricity.

wave-cut platform *noun*
A wave-cut platform is a flat shelf of rock found at the base of **cliffs** on the **shore**. It is all that is left of cliffs which have been **eroded** by **waves**.
As the cliffs were eroded further, the wave-cut platform became wider and wider.

weather forecast *noun*

A weather forecast is a statement how the weather is likely to behave in the future. Ships at **sea** need to know how the weather will change in the next few hours. In mist and **fog**, they must avoid going too close to rocks or **reefs**. If strong **winds** are forecast, sailors must prepare the ship for rough seas. Small boats avoid setting out if storms or **gales** are forecast.
The yachtsmen heard on the weather forecast that bad weather was coming and returned to shore.

Weddell Sea *noun*

The Weddell Sea is an arm of the **Antarctic Ocean**. It is found between the Antarctic Peninsula and Coats Land in Antarctica. Much of the Weddell Sea is covered by **ice shelves** and **pack ice**. Several research bases lie along the shores of the Weddell Sea.
Melting ice that comes from the Weddell Sea feeds many of the world's deep ocean currents.

weeverfish *noun*

A weeverfish is a small **fish** which has poisonous spines on the front part of the dorsal **fin** and on its **gill** covers. Weeverfish are usually found in European coastal waters. They can measure up to 40 centimetres in length. Weeverfish lie half-buried in **sand** in shallow water, searching for **prey** with their upward-facing eyes. They pose a threat to bathers and paddlers, and deliver a painful sting which can lead to serious infections.
The girl noticed the black spines of a weeverfish sticking out of the sand.

West Australian Current *noun*

The West Australian Current is a cold **surface current** in the **Indian Ocean**. It branches off from the **West Wind Drift** and flows up the west **coast** of Australia to join the south **Equatorial Current**.
The West Australian Current is weak in winter but strong in summer.

West Wind Drift *noun*

The West Wind Drift is a large, cold current which flows west around Antarctica, mainly between **latitudes** 40 and 60 degrees south.
The West Wind Drift is also called the Antarctic Circumpolar Current.

Westerlies *plural noun*

The Westerlies, or Westerly Winds, are the **prevailing winds** between the **Trade Wind** belt and **polar** regions. In the **northern hemisphere** they blow from the south-west and in the **southern hemisphere** they blow from the north-west.
In the North Atlantic, the Westerlies cause the Gulf Stream to curve towards Europe.

whale ► page 156

whale song *noun*

Whale song describes a sequence of sounds made by certain **whales** under water, especially during the breeding season. These songs can be heard hundreds of kilometres away in the **ocean**.
No one knows what the meaning of whale songs is.

whalebone ► baleen

whaling *noun*

Whaling describes the catching and killing of **whales**. Whale-catchers use large ships called whalers to follow the whales.
A **harpoon** gun is mounted on the front of the ship ready to fire. The dead whales are cut up and processed on board the whaler.
Whaling has drastically reduced the numbers of whales in the oceans.

whale *noun*

A whale is a large marine mammal. There are about 75 **species** of whale divided into two main groups. The toothed whales include **killer whales**, pilot whales and **narwhals**. They feed mainly on **squid** and fish. All toothed whales swallow their prey whole. The baleen whales include the blue whale and the humpback whale. They feed by filtering **zooplankton** from the water. *The humpback whale came up to the surface of the water for air.*

humpback whale up to 15 metres long

sei whale up to 17 metres long

right whale up to 18 metres long

sperm whale up to 20 metres long

fin whale up to 24 metres long

blue whale up to 30 metres long

Whales cannot breathe through their mouth. Instead, they breathe through their nostrils which form the blowhole on top of their head. A thick, muscular flap of skin keeps the water out when the whale submerges.

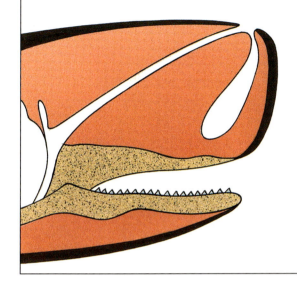

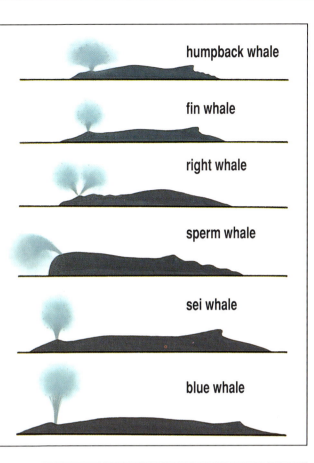

humpback whale

fin whale

right whale

sperm whale

sei whale

blue whale

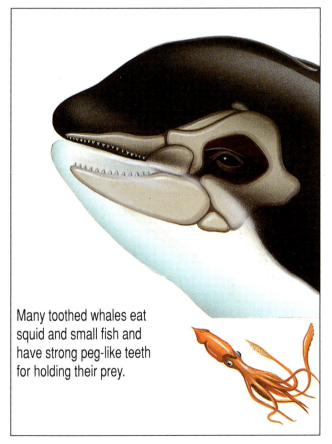

Many toothed whales eat squid and small fish and have strong peg-like teeth for holding their prey.

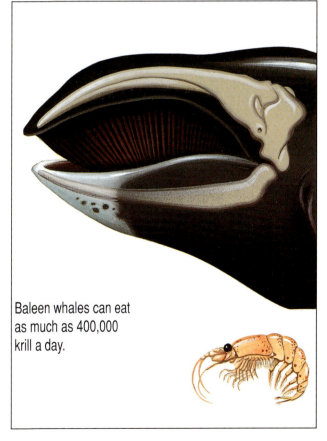

Baleen whales can eat as much as 400,000 krill a day.

whelk *noun*
A whelk is a **gastropod** mollusc with a ridged, spiral shell. It is found on the seashore, where it feeds on echinoderms and other **molluscs**, especially **bivalves**. Whelks have a long proboscis, or tongue-like structure, which can be stretched out or pulled in, with hard, horny teeth which can bore through the shells of their **prey**.
At the seaside there are stalls which sell whelks to eat.

whirlpool *noun*
A whirlpool is a violent **eddy** in the **sea**. It is formed where rising and falling **tides** meet. In some whirlpools, the water **currents** spiral down into the **ocean**. Smaller whirlpools form in rivers when rapidly flowing water meets an obstacle.
The men rowing the boats were afraid of being sucked into the whirlpool.

White Sea *noun*
The White Sea is an arm of the **Arctic Ocean**. It is almost completely surrounded by the Kola Peninsula and the north-east **coast** of Russia. The White Sea is linked to the **Barents Sea** by a long strait called the Gorlo. The White Sea-Baltic Canal links the White Sea to the **Gulf of Finland** at Leningrad. The White Sea is covered in ice between September and June. It is used heavily for shipping in the summer.
The Onega, Duina and Mezen rivers flow into the White Sea.

whitebait *noun*
A whitebait is a small, silvery **fish**. Whitebait is the young of **herrings**, sprats and silversides. It is eaten as a delicacy.
Whitebait are about 10-15 centimetres long.

whiting ► **cod family**

wind *noun*
Wind is movement of the air. Wind moves from **high pressure areas** to **low pressure areas**. The direction of the wind is affected by the **Coriolis force**, the shape of the land and by **depressions** and **anticyclones**.
Cold, icy winds blow down from the icefields of Antarctica.

wind surfing *noun*
Wind surfing is a sport similar to **surfing**, in which the surf-board has a sail stretched across a bar. The surfer stands on the board and holds on to the sail bar. The board is driven by the **wind**, not the **waves**, so wind surfing can be enjoyed on lakes and in sheltered **bays**.
There was enough breeze for wind surfing.

wobbegong *noun*
A wobbegong is a large ray-like **shark** found in shallow water around the **coasts** of Australia. It grows up to three metres in length, and has a flattened, rounded body with a tapering tail. The wobbegong has an attractive pattern of light-edged spots and bars.
In caves and hollows on the Barrier Reef divers may come across wobbegongs.

worm *noun*

A marine worm is a long, cylinder-shaped **invertebrate** which lives on or in the sea-bed, and in the **intertidal zone**. There are many different kinds of marine worm. The body of an annelid worm is divided into sections. Scaleworms are scaly, while the **sea mouse** has a bristly coat. Ragworms swim with tiny paddles, but flatworms flap their wide body up and down. Tubeworms build tubes of sand to live in. They **filter-feed** with feathery **tentacles**. Proboscis worms collect **detritus** on a sticky tongue. Lugworms eat their way through the **sediment**.

Marine worms have many different life-styles, from predator to scavenger.

worm cast *noun*

A worm cast is a small mound of **sand** or **mud** shaped like a corkscrew. It is formed by marine **worms** such as lugworms. These worms eat their way through the **sediment**, breaking down, or digesting, any **detritus** it contains and pushing out the rest as casts.

The sand exposed at low tide was covered in worm casts.

wrack *noun*

A wrack is a **brown seaweed**. Wracks have flat, branching leaves or fronds, covered with air bladders to keep them afloat. They are very common on rocky shores. There are several different types of wracks, including spiral wracks, channelled wracks and knotted wracks.

Different wracks grow at different levels on the shore.

wrasse *noun*

A wrasse is a colourful fish which lives in shallow water. Wrasses have a small mouth and large teeth which sometimes stick out between their lips. Most wrasses are **carnivores**. Many have extra grinding teeth in their throat for crushing **molluscs**.

Some female wrasses change into males, altering their colour at the same time.

wreckfish *noun*

The wreckfish is a fish that gets its name from its habit of visiting wrecks for food, or drifting alongside floating timber. Wreckfish are bluish-grey in colour. They can reach a length of 1.83 metres and weigh about 32 kilograms.

Wreckfish are found in the Atlantic Ocean.

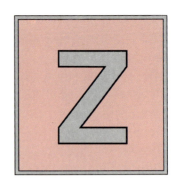

Yellow Sea   *noun*
The Yellow Sea is a large, shallow arm of the **Pacific Ocean**, between the Korean Peninsula and the **coast** of north-east China.
The Yellow River of China carries yellow silt into the Yellow Sea.

zooplankton   *noun*
Zooplankton describes the animals in **plankton**, such as **jellyfish**, water fleas and **copepods**.
Many small fish feed on the zooplankton.

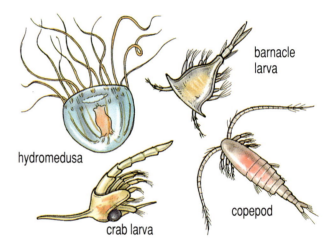

Zuider Zee   *noun*
The Zuider Zee is an inlet of the **North Sea**. It is found in the Netherlands between the West Frisian Islands and the Dutch mainland. A long dam now divides the Zuider Zee into the outer Waddenzee and the inner Ijsselmeer. The Ijsselmeer has become a freshwater lake, and much of it has been turned into **polders**.
Many waders feed on the shores of the Zuider Zee.